Oxford Series in Ecology and Evolution
Edited by Paul H. Harvey, Robert M. May, H. Charles J. Godfray, and Jennifer A. Dunne

The Comparative Method in Evolutionary Biology
Paul H. Harvey and Mark D. Pagel
The Cause of Molecular Evolution
John H. Gillespie
Dunnock Behaviour and Social Evolution
N. B. Davies
Natural Selection: Domains, Levels, and Challenges
George C. Williams
Behaviour and Social Evolution of Wasps: The Communal Aggregation Hypothesis
Yosiaki Itô
Life History Invariants: Some Explorations of Symmetry in Evolutionary Ecology
Eric L. Charnov
Quantitative Ecology and the Brown Trout
J. M. Elliott
Sexual Selection and the Barn Swallow
Anders Pape Møller
Ecology and Evolution in Anoxic Worlds
Tom Fenchel and Bland J. Finlay
Anolis Lizards of the Caribbean: Ecology, Evolution, and Plate Tectonics
Jonathan Roughgarden
From Individual Behaviour to Population Ecology
William J. Sutherland
Evolution of Social Insect Colonies: Sex Allocation and Kin Selection
Ross H. Crozier and Pekka Pamilo
Biological Invasions: Theory and Practice
Nanako Shigesada and Kohkichi Kawasaki
Cooperation Among Animals: An Evolutionary Perspective
Lee Alan Dugatkin
Natural Hybridization and Evolution
Michael L. Arnold
The Evolution of Sibling Rivalry
Douglas W. Mock and Geoffrey A. Parker
Asymmetry, Developmental Stability, and Evolution
Anders Pape Møller and John P. Swaddle
Metapopulation Ecology
Ilkka Hanski
Dynamic State Variable Models in Ecology: Methods and Applications
Colin W. Clark and Marc Mangel
The Origin, Expansion, and Demise of Plant Species
Donald A. Levin
The Spatial and Temporal Dynamics of Host-Parasitoid Interactions
Michael P. Hassell
The Ecology of Adaptive Radiation
Dolph Schluter

Parasites and the Behavior of Animals
Janice Moore
Evolutionary Ecology of Birds
Peter Bennett and Ian Owens
The Role of Chromosomal Change in Plant Evolution
Donald A. Levin
Living in Groups
Jens Krause and Graeme D. Ruxton
Stochastic Population Dynamics in Ecology and Conservation
Russell Lande, Steiner Engen and Bernt-Erik Sæther
The Structure and Dynamics of Geographic Ranges
Kevin J. Gaston
Animal Signals
John Maynard Smith and David Harper
Evolutionary Ecology: The Trinidadian Guppy
Anne E. Magurran
Infectious Diseases in Primates: Behavior, Ecology, and Evolution
Charles L. Nunn and Sonia Altizer
Computational Molecular Evolution
Ziheng Yang
The Evolution and Emergence of RNA Viruses
Edward C. Holmes
Aboveground–Belowground Linkages: Biotic Interactions, Ecosystem Processes, and Global Change
Richard D. Bardgett and David A. Wardle
Principles of Social Evolution
Andrew F. G. Bourke
Maximum Entropy and Ecology: A Theory of Abundance, Distribution, and Energetics
John Harte

Maximum Entropy and Ecology

A Theory of Abundance, Distribution, and Energetics

JOHN HARTE
University of California Berkeley, CA 94720 USA

UNIVERSITY PRESS

OXFORD
UNIVERSITY PRESS

Great Clarendon Street, Oxford ox2 6dp

Oxford University Press is a department of the University of Oxford.
It furthers the University's objective of excellence in research, scholarship,
and education by publishing worldwide in

Oxford New York

Auckland Cape Town Dar es Salaam Hong Kong Karachi
Kuala Lumpur Madrid Melbourne Mexico City Nairobi
New Delhi Shanghai Taipei Toronto

With offices in

Argentina Austria Brazil Chile Czech Republic France Greece
Guatemala Hungary Italy Japan Poland Portugal Singapore
South Korea Switzerland Thailand Turkey Ukraine Vietnam

Oxford is a registered trade mark of Oxford University Press
in the UK and in certain other countries

Published in the United States
by Oxford University Press Inc., New York

© John Harte 2011

The moral rights of the author have been asserted
Database right Oxford University Press (maker)

First published 2011

All rights reserved. No part of this publication may be reproduced,
stored in a retrieval system, or transmitted, in any form or by any means,
without the prior permission in writing of Oxford University Press,
or as expressly permitted by law, or under terms agreed with the appropriate
reprographics rights organization. Enquiries concerning reproduction
outside the scope of the above should be sent to the Rights Department,
Oxford University Press, at the address above

You must not circulate this book in any other binding or cover
and you must impose the same condition on any acquirer

British Library Cataloguing in Publication Data

Data available

Library of Congress Cataloging in Publication Data

Data available

Typeset by SPI Publisher Services, Pondicherry, India
Printed in Great Britain
on acid-free paper by
CPI Antony Rowe, Chippenham, Wiltshire

ISBN 978–0–19–959341–5 (Hbk)
       978–0–19–959342–2 (Pbk)

1 3 5 7 9 10 8 6 4 2

To the wild birds,
whose mere shadows are captured here

Preface

This book builds the foundation for, and constructs upon it, a theory of ecology designed to explain a lot from a little. By "little" I mean only inferential logic derived from information theory. And by "a lot" I mean the ecological phenomena that are the focus of macroecology: patterns in the partitioning of space and energy by individual organisms and by species in ecosystems.

Parts I–III are foundational. Part I is epistemological, describing the roles and types of theory in ecology and setting forth the concept of the logic of inference. I give special attention here to varying views concerning the role of mechanism in science. Part II is a self-contained primer on macroecology. I provide an extensive review of the field, covering the questions asked, the metrics used, the patterns observed to prevail, applications in conservation biology, and the broad categories and examples of theories advanced to explain those patterns. In Part III, I teach the maximum entropy (MaxEnt) method of inference from information theory, clarify the connection between information entropy and thermodynamic entropy, and discuss previous applications of the maximum entropy method in fields other than ecology.

In Part IV, I construct the MaxEnt Theory of Ecology (METE),[1] derive its predictions, and compare them to empirical patterns. In Part V, I draw connections between MaxEnt and other prominent ecological theories such as metabolic theory and the neutral theory of ecology. I also review other approaches to applying MaxEnt in ecology and potential applications of METE to problems in conservation biology. In the last chapter of Part V, I describe "loose ends" and possible ways to tie them up. METE is a static theory in which patterns in the abundance, distribution, and energetics of species in an ecosystem at any given moment in time derive from the instantaneous values of a small set of quantities, called state variables. As a possible extension of METE, I propose a way to construct a dynamic version, in which the time dependence of state variables can be predicted. Other unresolved issues are described and a set of research topics are suggested that I hope inspire some readers to engage in the effort to further advance the theory.

---

[1] Mete: (mēte) vt: to measure out appropriately, allocate.

My primary aim is a textbook that will provide readers with the concepts and practical tools required to understand (and contribute to) a rising area of science, generally, and macroecology in particular: the MaxEnt method. While the book's topical focus is a theory of macroecology based on the concept of maximum information entropy, I will evaluate the success of the book by the number of its readers who become empowered to use the methods described in it for any purpose whatsoever.

Exercises for the reader are provided at the end of each chapter. More difficult ones are noted with a * or a * *, with * * reserved for problems of term-paper magnitude or beyond. Some of the exercises involve just filling in steps in derivations or are qualitative discussion topic suggestions.

Used as a textbook for a one semester graduate seminar or lecture course covering topics in theoretical ecology, macroecology, and maximum entropy methods, instructors might find it reasonable to have students read and discuss Part I for the first week of the course. Because Part II introduces many quantitative concepts and methods that may be unfamiliar to students, it will probably require four weeks for adequate coverage of the material, two weeks for each of Chapters 3 and 4. Part III can be covered in two more weeks—one week to deal with the concepts in Chapter 5 and one to go over examples of MaxEnt calculations as described in Chapter 6. Three weeks on Part IV should be sufficient, with most of that time devoted to Chapter 7. Part V could be covered in as little as two weeks, assuming the material in Chapter 11 is not used to launch student research projects. If the class chooses to conduct term projects based on the suggestions in Chapter 11, an additional two or three weeks could be set aside for that, adding up to a 14 or 15 week course.

By way of warning, the subject matter herein contains material of a mathematical nature. To allow the reader to exercise discretion, truly $f(x)$-rated material, such as difficult derivations, is set aside in boxes and Appendices for the mathematically prurient. Their ecological content is explained in the main text, so that reading and understanding them is not essential to understanding concepts, or to being able to use the theory. To assist users, I explain, and provide explicit directions for using, a variety of computational tools to work out and compare with data the predictions of theory.

Two sources of funding, the US National Science Foundation and the Miller Foundation, provided me with the opportunity to develop many of the ideas herein. Much of the material is based upon work supported by the National Science Foundation under Grant No. 0516161.

I am indebted to numerous people for inspiration, technical advice, and occasional harsh criticism (without which I would have included material here that I would surely come to regret). First, I thank my current and former graduate students who have worked with me in the development and testing of ecological theory; they include Danielle Christiansen, Erin Conlisk, Jennifer Dunne, Jessica Green, Ann Kinzig, Justin Kitzes, Neo Martinez, Erica Newman, Annette Ostling, and Adam Smith. Two postdocs made invaluable contributions as well: Uthara Srinivasan and

Tommaso Zillio. Others, whose insights have hugely enriched my understanding of macroecology or maximum entropy, include Drew Allen, Jayanth Banavar, Luis-Borda de Agua, Jim Brown, Rick Condit, John Conlisk, Fred Cummings, Roderick Dewar, Brian Enquist, Jim Gillooly, Patty Gowaty, Ken Harte, Fangliang He, Stephen Hubbell, R. Krishnamani, Bill Kunin, Amos Maritan, Pablo Marquet, Brian McGill, Shahid Naeem, Jeff Nekola, Graham Pyke, Mark Ritchie, Michael Rosenzweig, Jalel Sager, Ida Sognnaes, David Storch, Geoffrey West, and Ethan White. Many people associated with Oxford University Press have been critical to the production of this book. Ian Sherman, Jennifer Dunne, Helen Eaton, Muhammed Ridwaan, Abhirami Ravikumar and Jeannie Labno deserve special thanks for their vision, effort, and skill.

While I cannot thank them all, there are thousands of professional ecologists, indigenous people, and volunteers who have over the years made this work possible. They are the census-takers, the counters of plants and animals who have laboriously set up gridded landscapes or walked transects, scoured the ground or the tree tops, watched and listened, pored through taxonomic keys, and entered data that I have used to test theory. I am quite sure that the rewards they obtained from close encounters with the living treasures of our planet more than make up for the absence of anything but a generic thanks from me.

Both my daughter, Julia, and my wife, Mel, have been patient and critically responsive to my Ancient-Mariner-like efforts over the past few years to explain maximum entropy to them, thereby allowing me to understand it better. The love and support my family has provided is immeasurable.

Finally, while I never had the honor to meet Edwin Jaynes, he probably deserves the most thanks of all. His papers developing the concept of maximum information entropy are brilliant; during the course of writing this book there were many times I found myself wishing I could talk to him in person.

Contents

Preface vi
Glossary of symbols xiv
Acronyms and abbreviations xv

Part I Foundations

1 The nature of theory — 3

1.1 What is a theory? — 3
   1.1.1 Falsifiability — 3
   1.1.2 Comprehensiveness — 4
   1.1.3 Parsimony — 4
   1.1.4 Easier said than done — 4
1.2 Ecology and physics — 6
1.3 Types of theories — 8
   1.3.1 The role of mechanism in science — 9
   1.3.2 Mechanistic theories and models in ecology — 11
   1.3.3 Statistical theory and models in ecology — 12
   1.3.4 Neutral theories of ecology — 12
   1.3.5 Theories based on an optimization principle — 13
   1.3.6 State variable theories — 14
   1.3.7 Scaling theories — 14
1.4 Why keep theory simple? — 16
1.5 Exercises — 17

2 The logic of inference — 19

2.1 Expanding prior knowledge — 20
2.2 Sought knowledge can often be cast in the form of unknown probability distributions — 21

x • Contents

2.3 Prior knowledge often constrains the sought-after distributions 22
2.4 We always seek the least-biased distribution 22
2.5 Exercises 23

Part II Macroecology

3 Scaling metrics and macroecology 27

3.1 Thinking like a macroecologist 27
    3.1.1 Questioning like a macroecologist 27
    3.1.2 Censusing like a macroecologist 28
3.2 Metrics for the macroecologist 29
    3.2.1 Units of analysis 30
3.3 The meaning of the metrics 32
    3.3.1 Species-level spatial abundance distribution 33
    3.3.2 Range–area relationship 35
    3.3.3 Species-level commonality 37
    3.3.4 Intra-specific energy distribution 39
    3.3.5 Dispersal distributions 40
    3.3.6 The species–abundance distribution (SAD) 41
    3.3.7 Species–area relationship (SAR) 41
    3.3.8 The endemics–area relationship (EAR) 46
    3.3.9 Community commonality 46
    3.3.10 Community energy distribution 47
    3.3.11 Energy– and mass–abundance relationships 47
    3.3.12 Link distribution in a species network 51
    3.3.13 Two other metrics: The inter-specific dispersal–abundance relationship and the metabolic scaling rule 52
3.4 Graphs and patterns 52
    3.4.1 Species-level spatial-abundance distributions: $\Pi(n|A, n_0, A_0)$ 61
    3.4.2 Range–area relationship: $B(A|n_0, A_0)$ 63
        3.4.2.1 A note on the nomenclature of curvature 63
    3.4.3 Species-level commonality: $C(A,D|n_0, A_0)$ 63
    3.4.4 Intra-specific distribution of metabolic rates: $\Theta(\varepsilon|n_0)$ 65
    3.4.5 Intra-specific distribution of dispersal distances: $\Delta(D)$ 65
    3.4.6 The species–abundance distribution: $\Phi(n|S_0, N_0, A_0)$ 66
    3.4.7 The species–area relationship: $\bar{S}(A|N_0, S_0, A_0)$ 68
    3.4.8 The endemics–area relationship: $\bar{E}(A|N_0, S_0, A_0)$ 73
    3.4.9 Community-level commonality: $X(A,D|N_0, S_0, A_0)$ 73
    3.4.10 Energy and mass distributions and energy– and mass–abundance relationships 75
    3.4.11 $\Lambda(l|S_0, L_0)$ 76
    3.4.12 $\varepsilon(m)$ 77

3.5	Why do we care about the metrics?	78
	3.5.1 Estimating biodiversity in large areas	78
	3.5.2 Estimating extinction	79
	3.5.3 Estimating abundance from sparse data	82
3.6	Exercises	83

4 Overview of macroecological models and theories   87

4.1	Purely statistical models	87
	4.1.1 The Coleman model: Distinguishable individuals	87
	4.1.2 Models of Indistinguishable Individuals	89
	4.1.2.1 Generalized Laplace model	92
	4.1.2.2 HEAP	93
	4.1.3 The negative binomial distribution	95
	4.1.4 The Poisson cluster model	97
4.2	Power-law models	97
	4.2.1 Commonality under self-similarity	100
4.3	Other theories of the SAD and/or the SAR	102
	4.3.1 Preston's theory	102
	4.3.2 Hubbell's neutral theory of ecology	104
	4.3.3 Niche-based models	105
	4.3.4 Island biogeographic theory	108
4.4	Energy and mass distributions	109
4.5	Mass–abundance and energy–abundance relationships	110
4.6	Food web models	110
4.7	A note on confidence intervals for testing model goodness	111
4.8	Exercises	112

Part III The maximum entropy principle

5 Entropy, information, and the concept of maximum entropy   117

5.1	Thermodynamic entropy	117
5.2	Information theory and information entropy	121
5.3	MaxEnt	123

6 MaxEnt at work   130

6.1	What if MaxEnt doesn't work?	130
6.2	Some examples of constraints and distributions	131
6.3	Uses of MaxEnt	133
	6.3.1 Image resolution	133

	6.3.2	Climate envelopes	133
	6.3.3	Economics	134
	6.3.4	Food webs and other networks	135
	6.3.5	Classical and non-equilibrium thermodynamics and mechanics	135
	6.3.6	Macroecology	136
6.4	Exercises	136

Part IV Macroecology and MaxEnt

7 The maximum entropy theory of ecology (METE) 141

7.1	The entities and the state variables	141
7.2	The structure of METE	142
	7.2.1 Abundance and energy distributions	142
	7.2.2 Species-level spatial distributions across multiple scales	146
7.3	Solutions: $R(n, \varepsilon)$ and the metrics derived from it	146
	7.3.1 Rank distributions for $\Psi(\varepsilon)$, $\Theta(\varepsilon)$, and $\Phi(n)$	152
	7.3.2 Implications: extreme values of n and ε	153
	7.3.3 Predicted forms of other energy and mass metrics	155
7.4	Solutions: $\Pi(n)$ and the metrics derived from it	157
7.5	The predicted species–area relationship	162
	7.5.1 Predicting the SAR: Method 1	163
	7.5.2 The special case of $\bar{S}(A)$ for $1 - A/A_0 \ll 1$	166
7.6	The endemics–area relationship	167
7.7	The predicted collector's curve	168
7.8	When should energy-equivalence and the Damuth relationship hold?	169
7.9	Miscellaneous predictions	173
7.10	Summary of predictions	174
7.11	Exercises	175

8 Testing METE 177

8.1	A general perspective on theory evaluation	177
8.2	Datasets	178
	8.2.1 Some warnings regarding censusing procedures	180
8.3	The species-level spatial abundance distribution	180
	8.3.1 A note on use of an alternative entropy measure	186
8.4	The community-level species–abundance distribution	186
8.5	The species–area and endemics–area relationships	190
8.6	The distribution of metabolic rates	193
8.7	Patterns in the failures of METE	196
8.8	Exercises	197

Part V A wider perspective

9 Applications to conservation — 201

9.1 Scaling up species' richness — 201
9.2 Inferring abundance from presence–absence data — 202
9.3 Estimating extinction under habitat loss — 202
9.4 Inferring associations between habitat characteristics and species occurrence — 203
9.5 Exercises — 206

10 Connections to other theories — 208

10.1 METE and the Hubbell neutral theory — 208
10.2 METE and metabolic scaling theories — 209
10.3 METE and food web theory — 210
10.4 Other applications of MaxEnt in macroecology — 210
10.5 Exercise — 212

11 Future directions — 213

11.1 Incorporating spatial correlations into METE — 213
    11.1.1 Method 1: Correlations from consistency constraints — 213
    11.1.2 Method 2: A Bayesian approach to correlations — 216
11.2 Understanding the structure of food webs — 218
11.3 Toward a dynamic METE — 218
11.4 Exercises — 225

Epilogue: Is a comprehensive unified theory of ecology possible? What might it look like? — 229

Appendix A: Access to plant census data from a serpentine grassland — 232

Appendix B: A fractal model — 233

Appendix C: Predicting the SAR: An alternative approach — 240

References — 244

Index — 253

Glossary of symbols

For symbols used to designate macroecological metrics see Tables 3.1a, 3.1b, 3.2.

A_0:   the area of an ecosystem in which we have prior knowledge
A: the area in which we seek to infer information
N_0:   the total abundance of all species under consideration in an A_0
S_0:   the total number of species under consideration in an A_0
E_0:   the total rate of metabolic energy consumption by the N_0 individuals in an A_0
n_0:   the abundance of a selected species in an A_0
n: a variable describing the abundance of a species
n_{max} the abundance of the most abundant species in a community
r rank, as in a rank-ordered list of abundances
D: a geographic distance
L_0:   the total number of links in a network
l: a variable describing the number of links connected to a node (usually a species) in a network
x $e^{-\lambda_\Pi}$
ε: a variable describing the metabolic rate of an individual
$\bar{\varepsilon}$:   average metabolic rate of the individuals in a species
λ_1:   Lagrange multiplier corresponding to the constraint on $R(n, \varepsilon)$ imposed by average abundance per species
λ_2:   Lagrange multiplier corresponding to the constraint on $R(n, \varepsilon)$ imposed by average metabolic rate per species
λ_Π Lagrange multiplier corresponding to the constraint on $\Pi(n)$ imposed by average number of individuals per cell.
β: $\lambda_1 + \lambda_2$
σ: $\lambda_1 + E_0 \cdot \lambda_2$
γ: $\lambda_1 + \varepsilon \cdot \lambda_2$
σ_{SB} Stefan–Boltzmann constant (5.67×10^{-8} watts per m^2 per (degree K)^4

Acronyms and abbreviations

CCSR	cross-community scaling relationship
CLT	central limit theorem
EAR	endemics–area relationship
GMDR	global mass–density relationship
IMD	individual mass distribution
LMDR	local mass–density relationship
log	logarithm with base e (i.e. the natural log)
MAR	mass–abundance relationship
MaxEnt	maximum information entropy
METE	maximum entropy theory of ecology
MST	metabolic scaling theory
NBD	negative binomial distribution
NTE	neutral theory of ecology
pdf	probability density function
RPM	random placement model
SAD	species–abundance distribution
SAR	species–area relationship

Part I

Foundations

1

The nature of theory

This chapter addresses the question: What do we mean by a theory of ecology? To answer that question, we must first answer several others: What is a theory? What are the special challenges ecology poses for theory building and testing? Does theory have to be based on mechanisms to be scientific? What do we actually mean by "mechanism"? Can theories based on manifestly incorrect assumptions be of value in science? We will address these and other questions here, and emerge with a sharper idea of the nature of the theory to be developed in the subsequent chapters.

1.1 What is a theory?

The term "theory" is used by scientists to describe their most advanced systems of knowledge. Thus we have the theory of evolution, the atomic theory, the theory of relativity. This is quite different from the colloquial, sometimes foolish, use of the term, as in "I have a theory that climate scientists are engaged in a conspiracy to fool the public." Systems of knowledge deserve to be called scientific theories when three criteria are met.

1.1.1 Falsifiability

A theory can be tested; practical criteria exist by which the theory can be widely agreed to have been disproven. It is the clarity of its predictions, not necessarily the reasonableness of its assumptions, that allows falsification. An implication is that a scientific theory can never be proven. One can never perform all possible tests of a theory and thus there is always the possibility that a test will someday be performed that will contradict the predictions of even evolutionary theory, the atomic theory, or relativity. While scientists are not holding their breath in anticipation of that happening, the possibility that it could happen is the hallmark of science in general and scientific theory in particular. In a nutshell, theory must "stick its neck out"! Any theory builder is a risk taker because theory must be destructible.

A consequence of the above bears remembering if you encounter deniers of evolution or of the science of climate change: when you hear them say that evolution or global warming science are only theories and have never been proven, don't argue. Taken literally, they are correct. But you might help them overcome their misunderstanding of what science is about! Scientists can't prove their theories; they can only disprove, or improve, them.

1.1.2 Comprehensiveness

A theory predicts the answers to many different kinds of questions that can also be answered with experiments and observations. The predictions are applicable across a wide variety of conditions and phenomena. A successful theory, of course, provides sufficiently accurate predictions, not just any predictions!

Thus an equation that simply predicts when your pet finch will begin singing each morning of the year does not deserve to be called a scientific theory. Rather it is a model; and while a very good model can be as, or more, accurate than a very good theory, it has only limited scope.

In contrast, the theory of evolution appears to describe generational change in biota across all taxonomic categories, across all of biological time, across all habitats. As far as we know, the theory of special relativity works across all of space and time; likewise for the atomic theory and the theory of thermodynamics.

1.1.3 Parsimony

A theory is lean; it does a lot with a little. A useful measure of the leanness of a theory is the ratio of the number of distinct testable predictions it makes to the number of assumptions that the theory is based on; that ratio should be large. The Ptolemaic theory of the solar system grew porky over time, while the Copernican theory was lean. A candidate theory that has to make nearly as many assumptions as it makes explanations, is not very impressive. Scientists sometimes call extra-lean theories "elegant." We shall see in Part IV that the theory we develop in this book, METE, is quite lean; but parsimony is more than an aesthetic issue. A theory with numerous tunable parameters is likely to readily fit available data and thus is difficult to falsify. Related to this, the many different combinations of parameters likely to yield good fits to data describing the past might each make different predictions about the future, leaving our understanding of the future no more predicted than it was without the theory.

1.1.4 Easier said than done

These criteria are easy to state, but are they easy to evaluate when a new theory is advanced? We would like to have a theory of ecology that provides, at the very least, a usefully accurate and predictive understanding of resource partitioning by species and individuals, in a wide variety of habitat types and taxonomic groups, across a wide and ecologically interesting range of spatial and temporal scales. Suppose an ecologist advances such a comprehensive theory of ecology and wishes to convince others that it is falsifiable and lean.

Consider falsifiability. How wrong does a prediction have to be before we decide the theory needs to be improved or discarded? In physics we might question the value of a theory if its predictions only start to differ from observation out at the fifth decimal place, provided of course that our observations are sufficiently precise and accurate. By contrast, and for reasons we discuss in Section 1.2, in ecology we might be satisfied with a theory if it predicts the dominant trends in the data, and if the scatter in data points around the predictions is small compared to the ranges over which the variables vary. But there is no hiding from the problem that there is no absolute or objective way to decide when to dump a theory.

How many phenomena does a proposed theory have to predict to warrant being called a comprehensive theory, not simply a model of a particular process? One answer is "you know it when you see it...." Though unsatisfying, that may be the best we can do in applying this criterion of comprehisiveness. In macroecology, we might decide to reject the theory if it predicts abundance distributions reasonably well but not species–area relationships, or if it works for herbaceous plants but not trees. A further complication is that comprehensiveness sometimes comes at the expense of realism, resulting in a perplexing tradeoff. Which of two theories is better: one that predicts six things somewhat well, or one that predicts only two of those things but much more accurately?

Parsimony is arguably more susceptible to quantitative evaluation than is comprehensiveness. This notion of leanness is most obvious when one evaluates models. If the model that accurately predicted when your finch will sing each morning of the year contained 365 adjustable parameters, one for each data point, you would not take it seriously. If it contained no adjustable parameters, it could be interesting; you might be motivated to test it on caged canaries, maybe even wild birds. More generally, there are statistical criteria (Akaike, 1974) for comparing the goodness of models that contain differing numbers of adjustable parameters and for which the accuracy of the model predictions differ. One expects to fit data better with more parameters to tune, but these techniques allow us to take this into account when we compare models. However, it can be more difficult to count *ad hoc* assumptions underlying a model or theory than it is to count the adjustable parameters they contain. While it is often the case that if it feels lean, it is lean, as with comprehensiveness and falsifiability there are no strict rules here. Later, when we examine theories of ecological phenomena, we will return to all three criteria and in greater depth discuss how we evaluate the degree to which they are met.

Returning to our chapter title, "The nature of theory," we have come a short way and already stumbled on some difficulties regarding evaluation of candidate theories. Further, a difference between ecology and physics was uncovered: the accuracy and precision of measurement. Let's pursue this cross-science comparison and highlight more broadly some similarities and contrasts between ecology and physics, where the notion of a "Grand Unified Theory" is widely entertained.

1.2 Ecology and physics

It is the late-sixteenth century and you are a graduate student in Pisa, Italy, working as a research assistant for Galileo. He has asked you to obtain data using a gravity linear accelerator, otherwise known as the leaning tower. You are to drop objects of various densities, sizes, and shapes from the top of the tower so that you and your fellow student on the ground below can accurately measure and record their descent times. You find that different objects fall at different rates because air resistance has slowed some objects more than others. Shortly thereafter, you discover that when Galileo tells his colleagues about your findings he asserts that the objects released at the same time all reached the ground at the same time. You, of course, get very upset, and when his obvious distortion of your data is not caught by his peers, you are tempted to leave physics. Don't... at least not for that reason.

What Galileo inferred from the experiment was not information about real descent times but rather about an ideal situation, objects falling in a perfect vacuum. This is accepted procedure in physics, and Galileo's genius exhibited itself here as the capacity to extrapolate the actual messy observed findings to this idealized situation. By doing so he was able to formulate a marvelous idea: the principle of inertia; and that became a cornerstone of the Newtonian theory of gravity and motion.

Biologists, in contrast to physicists, often seem to be fascinated by the uniqueness of each of their objects of study. Many ecologists loathe the stripping away of detail, the blurring of distinctions. And justifiably so; after all, it is the differences between individuals belonging to the same species that drives natural selection and allows for complex social organization. It is the differences between species in an ecosystem that creates a myriad of ecological processes or functions, and generates patterns of great complexity and beauty.

This contrast amounts to a kind of cultural difference between the two sciences. Just as physicists make progress by imagining a perfect vacuum or a frictionless spring, biologists often operate by looking for exceptions and uniqueness. If a biologist's fascination is with a particular species, then what is special about that species is often what justifies lifelong study of it.

But biology is too complex a science to be characterized as simply as I have above, so let me now turn around and argue the opposite. Darwin (1859) famously wrote:

> *It is interesting to contemplate an entangled bank, clothed with many plants of many kinds, with birds singing on the bushes, with various insects flitting about, and with worms crawling through the damp earth, and to reflect that these elaborately constructed forms, so different in each other, and dependent on each other in so complex a manner, have all been produced by laws acting around us.*

This quote, in the concluding paragraph of *The Origin of Species*, is remarkable for the vision of unity under law that it invokes. Indeed, evolution is as grand and sweeping an example of scientific theory as one could ever hope to achieve. We don't require one theory for plants and another for animals, let alone a different one for each species. And yet Darwin was an astute observer of particularity, chronicling in his many books the exquisite details of nature, the unique adaptations found throughout the living world. In a real sense it was the study of complex and local detail that led Darwin to a theory stunningly simple and sweeping in scope.

This is the paradox of biology in general and of ecology in particular. Life that lacks the differences and details seen throughout nature cannot be imagined; homogeneous life probably could not even exist. Those details are what drive many of us to study the organisms around us. Yet despite all that, a comprehensive theory of biological change, one that can encompass the diversity of life, one that is no less grand and unified than the best of physical theory, was achieved by Darwin.

Nevertheless, differences between physics and ecology do exist, and not just in what we have called "cultural differences" in the two sciences. Developing and testing ecological theory poses certain problems that physics does not face, at least not to the same degree. The manifestations of physical law are certainly shaped by accidents, such as the chance co-occurrence of physical objects, but those accidents do not stand in the way of deciphering the basic laws. A Galileo can "see through" the accidents. Doing this in ecology is arguably harder for the following reasons:

- Unlike protons zipping around an accelerator, no two ecosystems are even nearly alike, and thus truly controlled experiments are impossible. Psychologists face the same problem, lacking pairs of identical people with whom to experiment. Cosmologists of course have it worse; they not only lack pairs of universes, they cannot even experiment with the one they have.
- Ecologists, like anthropologists and cultural linguists, are witnessing each year the progressive deterioration, and sometimes extinction, of the objects of their study. Land-use practices, invasions by exotic species, climate change, overharvesting, toxification, and other environmental stresses are changing the dominant patterns and processes in ecosystems. Imagine that physicists had to fear the impending extinction of the Higgs particle or of the phenomenon of high-temperature superconductivity. Cosmology is sometimes mentioned as being similar to ecology in that in both fields, scientists have to deal with unique events and inherently non-replicable systems (in contrast to, say, bench-top chemistry or accelerator physics), but imagine what cosmology would be like in a universe where people have been uncontrollably moving galactic clusters around from one region to another and are artificially speeding up the rate of stellar death!
- Accurate ecological measurements, such as a complete census of the organisms that live in an ecosystem, are impossible to obtain. Animals don't stay still and many try to avoid being seen; it is difficult to identify to species level the seedling stage of plants; and all attempts to infer with statistical methods the properties of a heterogeneous system from subsamples result in intrinsic uncertainty.
- Conducting controlled experiments on ecosystems at large spatial scales and over long time-periods is not feasible. Hence our knowledge of large-scale phenomena, such as effects of global climate change on biodiversity, is often based on

observed correlations over time or over spatial gradients. But then the "correlation does not imply causation" trap can bedevil efforts to identify causation. At relatively small spatial and temporal scales, controlled manipulation experiments are often carried out on ecosystems, and these have been very effective in identifying underlying causal mechanisms operating at those scales. Unfortunately the theory needed to scale the consequences of these mechanisms up from plots to whole biomes, and from years to centuries or millenia, has been lacking, as has our understanding of emergent mechanisms that may only operate at these larger scales.
- System boundaries are often ill-defined. Attempts to carve out spatially and temporally delimited closed systems founder because of the difficulty of drawing space–time boundaries large enough to result in a system that is only weakly subject to outside influences, yet small enough to characterize.

We cannot wish these recalcitrant properties of ecosystems away, but rather we must, and will in this book, develop theory that is adapted to them. The next stage, then, in our inquiry into the nature of theory in ecology is to examine the various types of theory and to understand both their limitations and the opportunities they offer.

1.3 Types of theories

A distinction is often made between theories based upon explicit mechanisms of causation versus theories based upon statistical or other seemingly non-mechanistic assumptions. Evolution is a mechanistic theory in which the mechanism is selection and hereditability of traits acting in concert. In a nutshell, stressful forces upon organisms that differ genetically select those individuals possessing genes that confer on the individual and its offspring the greatest capacity to reproduce under those stresses.

Consider next a statistical explanation of the observation that the heights of a large group of children in an age cohort are well described by a Gaussian (aka normal) distribution. Invocation of the central limit theorem (CLT) provides a statistical explanation; but the question remains as to why that theorem applies to this particular situation. The applicability of the theorem hinges on the assumption that each child's height is an outcome of a sum of random influences on growth. So are we not also dealing here with a mechanistic explanation, with the mechanism being the collection of additive influences on growth that allow the applicability of the CLT? If the influences on growth were multiplicative rather than additive, we might witness a lognormal distribution. Is it possible that all scientific explanation is ultimately mechanistic? Let us look more carefully at the concept of mechanism in scientific explanation, for it is not straightforward.

1.3.1 The role of mechanism in science

In everyday usage, we say that phenomenon A is explained by a mechanism when we have identified some other phenomenon, B, that causes, and therefore explains, A. The causal influence of B upon A is a mechanism. However, what is accepted by one investigator as an explanatory mechanism might not be accepted as such by another. Consider the example of a person with red hair. A biochemist might assert that the mechanism causing red hair is the pigment pheomelanin. To a geneticist that pigment is just an intermediate phenomenon, the reflection of a mechanism, while the true mechanism is the gene that regulates the pigments that determine hair color. For a sociologist, perhaps the mechanism is the social forces that brought together two parents with the right combination of genes. The evolutionary biologist might invoke a mechanism by which the presence of that gene either conferred fitness or was linked to some other gene that conferred fitness, while someone taking a longer view might argue that the mechanism is the cosmic ray (or whatever else) that created the red-hair mutation in the ancestral genome long ago.

Does the search for mechanism inevitably propel us into an infinite regress of explanations? Or can mechanism be a solid foundation for the ultimate goal of scientific theory-building?

Consider two of the best established theories in science: quantum mechanics and statistical mechanics. Surprisingly, and despite their names, these theories are not actually based on mechanisms in the usual sense of that term. Physicists have attempted over past decades to find a mechanism that explains the quantum nature of things. This attempt has taken bizarre forms, such as assuming there is a background "aether" comprised of tiny things that bump into the electrons and other particles of matter, jostling them and creating indeterminancy. While an aether can be rigged in such a way as to simulate in matter the behavior predicted by Heisenberg's uncertainty principle, and some other features of the quantum world, all of these efforts have ultimately failed to produce a consistent mechanistic foundation for quantum mechanics.

Similarly, thermodynamics and statistical mechanics are mechanism-less. Statistical arguments readily explain why the second law of thermodynamics works so well. In fact, it has been shown (Jaynes, 1957a, 1957b, 1963) that information theory in the form of MaxEnt provides a fundamental theoretical foundation for thermodynamics.

One of the consequences of thermodynamics is the ideal gas law: $PV = nRT$, where P is the pressure exerted by a gas, V is the volume it occupies, T is its temperature (in degrees Kelvin), n is the number of moles of the gas, and R is a universal constant of nature. And here we can see an interesting example of how mechanism does play a role in this science. At temperatures and pressures experienced under everyday conditions, the ideal gas law assumes no interactions among gas molecules other than that they behave as point-like objects that generate no internal friction and heat when they bounce off one another. While the ideal gas law works spectacularly well under ordinary conditions, a systematic violation of the law is observed at sufficiently extreme values of pressure or temperature. The reason is that when molecules are shoved very close to one another, they exert a force on each other that can be ignored under ordinary conditions. The force is a consequence of molecules being composed of both positive and negative charges. It is called a dipole–dipole force, which you experience if you hold two magnets near one another. When these short-range forces exert their influence on the motion of molecules, they induce corrections to the ideal gas law; the corrected formula works quite well under these extreme conditions.

So with thermodynamics we have a theory that is mechanism-less and works well under everyday conditions but breaks down at high pressure. The breakdown is accounted for by the existence of what would conventionally be called a mechanism, dipole–dipole forces, which come into play at high pressure. But even those forces have their explanation in quantum mechanics, which we have seen is mechanism-less.

If we pull the rug of mechanism out from under the feet of theory, what are we left with? The physicist John Archibald Wheeler posited the radical answer "its from bits," by which he meant that information (bits)—and not conventional mechanisms in the form of interacting things moving around in space and time—is the foundation of the physical world (its). There is a strong form of "its from bits," which in effect states that only bits exist, not its. More reasonable is a weaker form, which asserts that our knowledge of "its" derives from a theory of "bits." It is this weaker form that METE exemplifies.

An example of this weak form of "its from bits" is suggested by quantum theory. Over the past decades there has grown a level of comfort with the notion that in quantum mechanics the absolute square of the wave function, which is a measure of probability, describes our current state of knowledge of a physical system. Nothing is gained, clarity is lost, and the door that lets quantum paradoxes enter is opened, if we assert that the square of the wave function is somehow a measure of the system itself. The system is real of course. Causes have effects, reality is not just in our minds, and there is no room in science for solipsism. But the quantum theory can be interpreted as a theory about our state of knowledge of the probabilities of possible states of a quantum system. When measurements on a system are performed, the wave function changes because we have gained new information about the system from the measurments. Probabilities in quantum theory describe our state of knowledge, not some intrinsic property of quantum states. Similarly, as formulated by Jaynes, thermodynamics is foundationally derived by maximizing information entropy, which as we will see in Chapter 5, is a way of minimizing the extent to which we are incorporating into our inferences assumptions that we have no basis for making.

Putting such ideas and their relevance to ecology aside for now, let's review our progress so far. We started out asking what a theory of ecology might look like and ended up wandering into physics to confront another problem. Mechanistic explanations either lead to an infinite regress of mechanism within mechanism, or to mechanism-less theory, or perhaps to Wheeler's world with its information-theoretic foundation. What is evident is that as we plunge deeply into the physical sciences, we see mechanism disappear. Yet equally problematic issues arise with statistical theories; we cannot avoid asking about the nature of the processes governing the system that allow a particular statistical theory to be applicable. In fact, when a statistical theory does reliably predict observed patterns, it is natural to seek an underlying set of mechanisms that made the theory work. And when the predictions fail, it is equally natural to examine the pattern of failure and ask whether some mechanism can be invoked to explain the failure.

Lest we wander any further from our goal, we return now to ecology and examine the broad categories of theory that have been advanced.

1.3.2 Mechanistic theories and models in ecology

What ingredients might go into a mechanistic theory of the distribution and abundance of species? Such a theory might be shaped by the ways in which species and individuals partition biotic and abiotic resources and interact with each other in doing so. The theory could include any combination of the major processes that most ecologists agree operate in nature: predation, competition, mutualism, commensalism, birth, death, migration, dispersal, speciation, disease and resistance to disease, an assortment of reproductive strategies ranging from lekking behavior to broadcast insemination to male sparring to flower enticement of pollinators—as well as an equally wide range of organism behaviors and social dynamics unrelated to reproduction, including harvesting strategies and escape strategies, plant signaling, nutrient cycling, plant architectures to achieve structural goals, life-cycle adaptations to fluctuations in weather and resource supply, and the list goes on. These are among the mechanisms governing millions of species and probably billions more locally adapted populations with distinguishable trait differences, all on a background of climatically and chemically heterogenous conditions, themselves influenced by the organisms. Out of this complexity is generated the macroecological phenomena we observe in nature. And just in case that list of ingredients to mechanistic theory does not offer enough choice, we could even include trait differences among individuals within populations. It seems a tall order, indeed, to attempt to pick from this menu a defensible set of ingredients for a mechanistic model.

Mechanistic ecological models designed to answer a relatively small set of questions are often built around variations on the Lotka–Volterra equations (Volterra, 1926; Lotka, 1956) describing species' interactions. These equations are usually used to describe a small subset of the above processes—such as predation, competition, birth, and death—with space homogenized and intra-specific trait differences ignored. Lotka–Volterra models have been useful in elucidating aspects of stability and coexistence of diverse species, but they are a long way from constituting a theory of ecology.

An obvious problem with mechanistic theory in ecology is the plethora of options, as listed above. A theory that incorporates some selected subset of recognized mechanisms and phenomena in ecology would still necessarily contain numerous tunable parameters. Therefore it would be difficult to falsify the theory or gain confidence in its predictions. We could perhaps begin to whittle down the list of processes, retaining some minimal set that leads to falsifiable, yet still comprehensive, theory. Alternatively, we could start with no mechanisms and then add them in when the data compel doing so. This leads us naturally to the topic of statistical theory and models in ecology.

1.3.3 Statistical theory and models in ecology

Here is an example of a statistical model of pattern in the spatial distribution of individuals in a species: I posit that each individual is placed randomly in the ecosystem without regard for where the other individuals happen to be found (Coleman, 1981), leading to a Poisson distribution of individuals in space (discussed in Chapter 4). This random placement model is parsimonious, it provides answers to a number of questions about the spatial structure of ecosystems (and thus is verging on a theory), and it is falsifiable. Indeed, it is falsified because individuals within species are generally observed to be distributed in a more aggregated pattern in space than predicted by the Poisson distribution. Of course, there is no reason to restrict statistical models to those that lead to Poisson distributions. Other statistical rules could be used. The essence of a statistical model is that some rule is followed consistently; some distribution is drawn from randomly. We will return later to other examples that work better than the Coleman random placement model.

Fortunately we can envision ecological theories that do not neatly fit within the purely mechanistic versus purely statistical dichotomy. A brief review of such theories will advance us toward the goal of a theory of ecology.

1.3.4 Neutral theories of ecology

Here we have a class of theories in which we partition life into species (or some other categories), but we distinguish the species only by their names, not by their traits. And because traits do not matter in neutral theories, neither does the heterogeneity of the abiotic environment. What does matter in such theories are stochastic events that influence outcomes. A fine example is Stephen Hubbell's Neutral Theory of Ecology (NTE) (Hubbell, 2001), in which species do not interact with one another or with their environment in any explicit differentiated way. Only stochastic birth, death, migration, and/or dispersal, and speciation events occur, leading to different outcomes for the number of species and their differing abundances. So at the outset, species are not distinguished, but stochastic demographic rates result in differences in species' abundances. Such a theory can, in principle, explain why some of the species of trees or birds in the forest are very abundant, while others are very rare, but because neutral theory is blind to traits, it cannot explain why colorful birds, say, are rarer than drab birds, or predict which species will do best under climate change.

1.3.5 Theories based on an optimization principle

It would be extremely useful if ecologists could predict patterns in the distribution and abundance of species in an ecosystem simply from the assumption that those patterns optimize some quantity. Economists have their utility function to maximize, physicists have a quantity they call "action" to minimize. What do ecologists get?

Do we maximize productivity, or longevity, or number of species, or some measure of diversity, such as the Shannon–Weaver metric (Shannon and Weaver, 1949; Krebs, 1989)? There are many choices, just as we saw there were in our discussion of mechanistic ecological theory. To understand spatial patterns in species' distributions, would we maximize the distance between competitors or minimize the distance between mutualists? Do we maximally fill some sort of abiotic niche space, or do we invoke the concept of ecological engineers and optimize over the interactions that simultaneously create niches and sustain biodiversity? Do we maximize the water-use efficiency of plant communities, or instead their light-use efficiency? Do we maximize stability; if so, stability of what, and against what stress? And which of the many types of stability, such as constancy, resistance, resilience (Holling 1973), do we optimize for? Would we optimize some trait over evolutionary time-scales or just over the demographic time-scale of a few generations? Do we take into account co-evolution and thus take into account body sizes and running speeds of predators and prey? Do ecosystems minimize Gibbs' free energy or perhaps optimize some other physical thermodynamic property?

One answer, implied by evolutionary theory, is that reproductive fitness is what should be maximized. But in various combinations, all of the above candidates for optimization could contribute to fitness. This highlights the central problem with optimization principles in complex systems: such systems may be optimizing over conflicting functions. Hence prior knowledge of the tradeoffs among these functions is required, yet a fundamental theory of the weight functions that determine the outcome of tradeoffs in ecology is lacking.

Why don't physicists have to worry about analogous complexities? Are those factors, listed earlier in the chapter, that distinguish physics and ecology at an epistemological level so hugely important that they allow optimization principles to be useful in physics but not in ecology? I will try to convince you in this book that the answer is no. Recall the vexing issues surrounding the notion of mechanism in science. The candidates for optimization in ecology mentioned above are in fact processes, mechanisms or, more generally, biological or physical characteristics of an ecosystem. Yet the most far-reaching physical theories do not rest on extremum principles that involve optimizing particular physical processes. Could the same be true in ecology?

We shall see that METE is based upon an extremum principle, but what is optimized in METE is not an ecosystem process or characteristic. Rather it is something about our knowledge itself. In particular, METE derives from maximizing a certain measure of the degree to which our inferred knowledge of an ecosystem contains no unwarranted assumptions, no bias. And that is just what I meant when I mentioned the weak form of "its from bits." METE is based on the idea that our knowledge of ecological "its" derives from optimization theory applied to the "bits" of our ecological knowledge.

1.3.6 State variable theories

The theory developed in Part IV describes macroecological patterns in terms of a small set of quantities that characterize an ecosystem. We call these quantities "state variables" in conformity with the usage of that term in thermodynamics. The state variables for, say, a container of gas are the volume of space it occupies, the number of moles of gas in that volume, and either the pressure the gas exerts or its temperature. These are quantities that define the system and they cannot be predicted from first principles. In ecology we might choose as a state variable for an ecosystem the area, A_0, that it occupies, which is certainly not deducible from something more basic. A second state variable might be the number of individuals, N_0, within that area and within some grouping, perhaps a taxonomic category, such as plants or birds. Total number of individuals is, like area, a deserving state variable for it cannot be inferred from area alone; there are hectares of desert with far fewer individual plants than are found in hectares of grassland. Another plausible candidate for a state variable for an ecosystem is the total number of species, S_0, within the chosen category, for again we can find systems with roughly similar area and similar total abundance but very different species' richness. Perhaps another state variable is the total rate of metabolic energy consumption, E_0, by all the individuals in the category, for again we can examine systems with roughly similar area, abundance, and species' richness, but with very different total metabolic rates of the organisms.

One question inevitably arises in identifying state variables: how do we know when we have enough? A theory with as many state variables as there are individual organisms would not be interesting. Ultimately, the only (and not completely satisfying) answer we can provide is that the best state variable theory will be the one with the fewest state variables that still works well. And as we will discuss in Section 10.4, there are situations where the addition of seemingly plausible extra state variables actually decreases the accuracy of a theory's predictions.

1.3.7 Scaling theories

The term "scaling" refers to mathematical regularities in the way the value of a variable changes when measured over different spatial or temporal scales or over

Types of theories • 15

Table 1.1 Hypothesized power-law scaling relationships

Phenomenon	Hypothesized Scaling Law	
Species' richness (S) as function of censused area (A) (Arhennius, 1921; Rosenzweig, 1995; but see Harte et al., 2009)	$S \propto A^z$	$z \approx ¼$
Variance in species richness (σ_s^2) across cells of given area as function of mean species richness ($\bar{S}$) in those cells (Taylor, 1961)	$\sigma_s^2 \propto \bar{S}^a$	$2 \geq a \geq 1$
Basal metabolic rate (ϵ) of organisms as function of body mass (m) (Kleiber, 1932; West et al., 1997; Brown et al., 2004; but see Kolokotrones et al., 2010)	$\epsilon \propto m^b$	$b \approx ¾$
Average time interval (T) between extinction events of magnitude E(Sole et al., 1997; but see Kirchner and Weil, 1998)	$T \propto E^c$	$c \approx 2$
Length of coastline (L) as a function of length of measuring stick (l) (Mandelbrot, 1982); d = fractal dimension	$L \propto l^{1-d}$	$2 \geq d \geq 1$
Average time interval (T) between earthquakes of Richter magnitude r(Gutenberg and Richter, 1954)	$T \propto r^q$	$q \approx 2$
Number of trophic links (L) in a food web as a function of number of species (S) (Martinez, 1992)	$L \propto S^c$	$c \approx 2$

other different conditions or circumstances. Consider Table 1.1, showing widely studied examples of either actually observed, or claimed but not well-substantiated, scaling patterns.

Scaling patterns are fascinating in their own right because when they are found in ecology, hydrology, geology, economics, or other fields, they illuminate a certain kind of simplicity behind what might have been thought to be idiosyncratic phenomena. After all, why should the simple relationship between earthquake magnitude and frequency, shown in the table, hold? Moreover, it is often possible to screen candidate theories or models on the basis of which ones generate an observed scaling pattern.

The scaling relationships in Table 1.1 are all of the form of what is called power-law behavior. This refers to the fact that in each case, the variable of interest, the dependent variable, depends on another variable through a mathematical power function: $y \sim x^b$ where b is the power or exponent. Not all scaling relationships have to be, or are, of that simple form, and we shall see that a non-power-law scaling relationship for the species–area pattern is predicted by METE. Power-law behavior is of special interest because it can be indicative of a property called "scale invariance." Fractals typify scale invariant patterns. The basic concept of self similarity is discussed in qualitative ecological terms in Section 4.2, and in Appendix B the mathematical formulation and predictions of models of scale invariance in ecology are presented.

One widely discussed scaling hypothesis in ecology relates the metabolic rate of an organism to its mass. A major focus here has been the ¾ scaling rule, which asserts:

$$\text{metabolic rate} = \text{constant}(\text{body mass})^{3/4}. \tag{1.1}$$

This result was first proposed based solely on empirical patterns (Kleiber, 1932) but subsequently several approaches to deriving it from fundamental theory were set forth. Most prominent of these approaches is the work of West, Brown, and Enquist (West et al., 1997; Brown et al., 2006), the architects of the metabolic scaling theory (MST). This bold theory of bioenergetics was motivated originally by the assumption that the blood circulatory system of animals is scale-invariant over the scale range from the largest to the smallest blood vessels.

An extension of the theory encompasses the energetics of vascular plants (again based on an assumption of scale invariance of the plant's fluid delivery system) (Enquist et al. 1999; Enquist and Niklas 2001). For animals, the metabolic rate is taken to be the basal, or resting, metabolic rate, while for plants, the equivalent energy variable is usually taken to be the respiration rate, which can be expressed in energy units using the conversion factor: 1 kg(respired carbon) ~ 40 megajoules.

A further extension of MST introduced body or environmental temperature as an additional factor influencing metabolism (Gillooly et al., 2001). In that extension of the theory, the "constant" in Eq. 1.1, is no longer a constant but instead increases at a predicted rate with temperature. The predictions of MST also include a variety of physiological properties of organisms related to metabolism and even include speciation rates.

MST is a scaling theory in the sense that it predicts a power-law relation between metabolic rate and body mass. It can also be construed as an optimization theory in the sense that minimization of the work to supply resources through a scale invariant branching network was the basis for the original derivation by West, Brown, and Enquist of the ¾ power scaling behavior. For varying views on the validity and the role in ecology of the MST, see Dodds et al. (2001), Kolokotrones et al. (2010), and the published MacArthur lecture by James Brown (Brown et al., 2006); note also the many brief responses, comprising a forum on metabolic theory, that immediately follow the Brown et al. article.

1.4 Why keep theory simple?

The two theories that we briefly described above, the NTE and the MST, as well as the theory we advance here, METE, are examples of relatively simple theories. Simple theories are based on relatively few, clearly stated, unabashedly over-simplified, assumptions that can be used to make a comprehensive set of falsifiable predictions about the answers to a wide variety of questions. In the absence of theory, those questions might have been considered independent of one another. The theory provides hope of unifying the questions and answers under one coherent framework. Each of those questions might be addressed with a mechanistic model tailored to each question, but the collection of those models does not constitute a theory.

Simple theories are "straw men" in the following sense: simple ecological theories are destined to fail, to make predictions contradicted by observation because these theories are based on assumptions that are not strictly true. Thus the NTE neglects traits that distinguish species, in the face of our knowledge that such traits are biologically important. Many of the assumptions underlying the MST, such as— in the original derivation—the assumption of a scale-invariant circulatory system, are not strictly true. There is ample scatter in the data around the central prediction of the MST and deviations from observation in the predictions of the NTE are also documented (Condit et al., 2002; Clark and McLachlan, 2003). These shortcomings might be considered sufficient evidence to dismiss such theories entirely, but given the parsimony and breadth of both these theories, such dismissal would not promote progress in ecology.

In fact, the failures of simple theories like these can lead to progress in science, as, for example, with the failure of the Ideal Gas Law leading to improved understanding of dipole–dipole forces between molecules. The failures of simple theories permit us to identify the situations in which some more nuanced assumptions are needed. Failure is what drives science forward.

By examining the instances in which simple yet comprehensive theory fails to describe all the data, and in particular looking carefully at the patterns in the discrepancies, we can establish more firmly the existence of influential mechanisms in nature that explicitly violate the simple assumptions underlying the theory. The outcome may improve the theory or lead to the development of a whole new theoretical construct, but in either case greater insight into ecology is achieved.

I have been unabashedly defending simple theory here, arguing, for example, that theories based on the assumption of neutrality can be of value, even though neutrality is clearly violated by observed species' traits (see also McGill and Nekola, 2010). For balance I urge readers to examine what is probably the most forceful critique of neutral theories and, indeed, of so-called non-mechanistic theory more generally (Clark, 2009). As you read any discussion of what constitutes desirable ecological theory, consider the possibility that arguments of the relative merits of simple versus complex theories, or of statistical versus mechanistic theories, will only end when, as has happened in other sciences, empirical evidence provides a reason to have confidence in the predictions of some emergent theory.

1.5 Exercises

Exercise 1.1

With reference to Section 1.3.6 think of an example of two ecosystems, which have the same area, same species' richness, and same total abundance of individuals, but in which the total rate of metabolism of all the individuals is considerably different.

Exercise 1.2

Read Clark (2009) and McGill and Nekola (2010) and write a dialogue that might occur if the authors engaged in a debate that allowed each to respond to the views of the other.

2

The logic of inference

Here I hope to convince you of the merit in a set of ideas that trace back to the mathematician/physicist Laplace but reach their fullest expression with the formal development of information theory and then the work of Edwin Jaynes in the twentieth century. The core ideas are that science is, to a considerable extent, based on the process of inference, and that scientific inference is based on logic. Here I present the underlying logic. If you accept the argument here, then you will be ready to learn how to use the principle of maximum information entropy in Part III.

Here, in the form of four premises, to be dissected below, is an axiomatic foundation for a logic of scientific knowledge acquisition:

1. In science we begin with prior knowledge and seek to expand that knowledge.
2. Knowledge is nearly always probabilistic in nature, and thus the expanded knowledge we seek can often be expressed mathematically in the form of probability distributions.
3. Our prior knowledge can often be expressed in the form of constraints on those distributions.
4. To expand our knowledge, the probability distributions that we seek should be "least biased," in the precise sense that the distributions should not implicitly or explicitly result from any assumptions other than the information contained in our prior knowledge.

Before you decide whether you accept the four premises in this axiomatic formulation of the logic of science, let me warn you of the consequence of doing so. In Part III, I will show you that there is a rigorously proven mathematical procedure, called information entropy maximization that implements step 4 above. That is, it produces the least biased form of the probability distribution that you seek, subject to the constraint of the prior knowledge. Thus acceptance of the above assertions entails more than acceptance of some esoteric philosophy-of-science framework. It entails acceptance of a process of inference, the maximum entropy method, which leads to scientific knowledge and theory construction. I do not claim that the framework presented above is applicable to all scientific theory construction, which is why the word "often" appears in statements 2 and 3. But because it is the basis for developing a theory of marcroecology in Part IV, you want to be very sure that you understand the assertions and the logic of the argument. So let's examine the assertions in more detail, in part to understand better their limitations.

2.1 Expanding prior knowledge

Here's an example that arises in ecology. Your prior knowledge consists of the results of canopy fumigation experiments, in which several dozen tree canopies scattered around the Amazonian basin are fumigated and the insects that rain out are counted and assigned a species' label. Thus you know the number of arboreal insect species in each of several dozen representative tree canopies. From this prior knowledge you wish to infer how many species of arboreal insects inhabit the entire Amazonian Basin.

This example has echoes in the more general field of image reconstruction. A sparse subsample of insects in canopies is somewhat like a fuzzy MRI image in medicine or a fuzzy photograph in forensics. The sparse and incomplete image is the prior knowledge about the more accurate image that is sought. Image reconstruction is, in fact, a widely-used application of the maximum information entropy method.

A second example arises in the analysis of topological networks. A topological (or qualitative) network is a web of nodes and the links connecting them. Your prior knowledge might consist of knowledge of the total number of nodes and the total number of links. Network analysts might want to know the fraction of nodes that are linked to just 1 other node, to 2 other nodes, and so on. In an ecological context, the network might be a food web or a pollinator–host plant web.

Next consider the case of an incomplete demographic preference matrix. The rows in the matrix are labeled by some demographic or economic trait, such as ethnicity or income or some other variable describing a set of categories into which people can be placed. The columns are labeled by a different one of those traits. For example, suppose the rows are income brackets and the columns are political parties. Then a particular matrix element might be the number of people in one of the income brackets who vote for a particular political party. And row sums correspond to the total number of people in each income bracket, while column sums correspond to the total number of people voting for each party. Your prior knowledge might consist of the row and column sums, from which you might wish to infer the unmeasured matrix elements themselves.

That third example is referred to by political scientists as "The Ecological Inference Problem" (King, 1997); despite that name, social scientists appear to have been more interested in it than have ecologists; but there are ecological examples. Consider a food web and construct a matrix in which both rows and columns are labeled by the species. A topological or qualitative web would be described by matrix elements that are either 1, if the row species eats the column species, or 0 (if the row species does not eat the column species). But we might know more than this, and, in particular, we might know the total amount of food required per unit time by each species and the rate of food production by each species. In other words, we might know the row and column sums, just as in the income–politics example. Then the solution to the Ecological Inference Problem would yield an estimate of the amount of food each species obtains from each other species, which gives us much more information than does just knowing the distribution of number of links across nodes.

As a final example, suppose your prior knowledge of the biodiversity in an ecosystem is knowledge of the total number of species in some taxonomic category, as well as the sum of the number of individuals in all those species. From this you wish to infer the distribution of abundances across all those species.

Common sense, and the breadth of examples above, suggests the validity of the premise that science seeks to expand prior knowledge.

2.2 Sought knowledge can often be cast in the form of unknown probability distributions

Taking our first example, above, the quantity we seek is the number of arboreal insect species in the entire Amazon. But from the available prior information, it is evident that at best all we can arrive at is knowledge of the form of a probability distribution. The expectation value of that distribution is then our best estimate of the number of species. Hence steps 1 and 2 in the framework above apply. This is also the case for the second and third examples concerning food web links. Here we seek a probability function, $\Lambda(l)$, informing us of the probability that a node has l links to other nodes or a similar type distribution for the amount of energy passing from one node to another. In the fourth example, the knowledge we seek is the probability that a species has a specified number of individuals. So in all these cases, items 1 and 2 in the framework apply.

Some might argue that we have no business talking about the probability distribution for the number of arboreal insect species in the Amazon because there is only one Amazon and thus a predicted probability distribution can never be tested. This argument stems from a narrow frequentist interpretation of probability, which asserts that the notion of probability only makes sense when you can do repeated draws from a "population" and thereby determine the frequency or probability with which different outcomes obtain. The best antidote for anyone afflicted with this hobbling interpretation of probability is reading Jaynes (2003), who makes the convincing case that probability theory is a systematic and consistent way to think about inference and expectation, and only secondarily about the results of repeated sampling.

We see that there are plenty of examples in which an ecological problem can be posed in terms of seeking probability distributions. But can all scientific problems be posed in those terms? Probably not. In biology it would be at best very awkward to approach questions related to the origin of species in terms of unknown probability distributions, and thus the development of the theory of evolution does not appear to fit well into this particular logical framework. In other words, I certainly do not claim that all of science falls within this framework, but only that enough science does to warrant exploring its further application. Are there ecological questions of current interest whose answers cannot be usefully formulated in the language of probability distributions? Exercise 2.2 gives you a chance to think about that.

2.3 Prior knowledge often constrains the sought-after distributions

We will show in Part IV that in the examples above, and many other ecological examples as well, the prior knowledge constrains the sought-after distribution, with the constraint often taking the form of knowledge of moments of the distribution. Consider the example above involving the distribution of species' abundances in an ecosystem. The ratio of the two pieces of prior knowledge, number of individuals divided by the number of species, is the average abundance per species, and is thus the expectation value of the abundance distribution of the species.

Are there sought-after probability distributions in ecology for which our prior knowledge does not provide constraints? Exercise 2.3 addresses that.

2.4 We always seek the least-biased distribution

This premise is, of course, unassailable in science; the real issue has to do with how we define minimum bias. Our definition states that the answer we obtain is least biased if it obeys all the constraints that result from our prior knowledge and is not constrained by anything other than that prior knowledge. It would be hard to argue with that, but the devil may be in the details. For example, we could get into trouble applying this premise if our prior knowledge is incorrect, or if we actually possessed more prior knowledge than we had assumed. So in principle, this premise is perfectly acceptable, but in practice we have to be alert to the possibility that the probability distribution that we obtain using premise 4 is flawed because our assumed information, which we call prior knowledge, may not be correct or complete.

The logic of inference is powerful. From our four premises, along with a fundamental theorem in information theory and a mathematical technique called the Method of Lagrange Multipliers, we have the machinery in hand to not just infer least-biased probability distributions, but to actually build theory. The theorem and the technique will be explained in Part III.

The theory, METE, that we build on this foundation will incorporate the traits of four types of theory described in Chapter: it will be a neutral theory, it will be based on an optimization criterion (minimizing bias in our inferred knowledge), it will be based on state variables appropriate to ecology, and it will result in (non-power-law) scaling relationships. It will also be conceptually simple, although the maths will look complicated at first. The theory will also help us understand better where mechanistic explanation needs to play a role in ecology and where it does not.

So now go back again and make sure you are comfortable with the four premises. Acceptance of these premises in macroecology, along with straightforward application of some mathematics, buys you more than you might have thought you would be willing to accept. The exercises below give you a chance to explore the limits of applicability of the four premises.

2.5 Exercises

Exercise 2.1

Make a list of three questions, arising either in everyday life or in your research or from classes, that interest and puzzle you. Then see which ones appear to fit the framework above in the sense that you have prior knowledge, you seek to extend it and infer some probability distribution, the prior knowledge can be considered to constrain the distribution you seek, and the notion of least-biased inference is relevant.

Exercise 2.2

Describe a question of current interest in ecology that cannot be usefully formulated in the language of a sought-after probability distribution.

Exercise 2.3

Describe a question of current interest in ecology for which our prior knowledge is not of a form that constrains a sought-after probability distribution.

Part II

Macroecology

3

Scaling metrics and macroecology

Here you will learn what macroecologists do: the things they measure, the way they measure them, the way they graph their data, and the patterns they tend to see when they stare at the graphs. You will be introduced to the concept of macroecological metrics, which are the mathematical entities that provide a way to compare patterns predicted by theory with patterns in nature. And, lest you think macroecology is purely esoteric, we will conclude with a survey of the many ways in which the findings of macroecology are applied in conservation biology to problems such as extinction under habitat loss.

3.1 Thinking like a macroecologist

Let's take a balloon ride and look down at about 100 km^2 of a high mountain range in late spring or early summer. Peering down you begin to wonder whether the plants that comprise the green patches near the summits are also adapted to the valleys below; if not, then the ecosystems near the summits could be like islands, isolated from one another by a sea of forested valleys instead of water. Do the patterns of biodiversity on the peaks resemble those on real islands? Soon you notice the open meadows within the mostly forested area; they also appear like islands of meadow amidst a sea of trees. And now you see that the patches of trees near timberline also seem to grow in clusters, like islands of trees in a sea of high meadow. Is such clustering the way of nature? Your questions soon multiply and become more specific.

3.1.1 Questioning like a macroecologist

Here are more specific questions that you might have thought about while in the balloon:

How many species of plants live above timberline on all the mountains in the entire region that is visible below you?
Do the flat-topped mountains, with more alpine area, contain more alpine species than the steeper ones wth less alpine area?
Some peaks are close to others while some may be quite distant from their nearest neighbors; are the relatively isolated summits more species' depauperate than the others? Do they contain rarer species?

Does the degree of overlap of the species' lists on two different summits depend primarily on the geographic distance between peaks, on the differences in their elevations, or on something else?

Now the balloon descends and you land on the flank of one of those mountains, just below its summit. Once on the ground you look around and even more questions arise:

Some plant species found in the alpine habitat where you landed appear rare, while others appear abundant. What is the distribution, over the species, of abundances? Would it be the same on all the summits? Would it be the same down below in the subalpine meadows that you see scattered through the forests there?

Similarly, some plant species seem to have quite small individuals, while others are larger. What is the distribution of sizes over all the species? Is it the same in different habitats?

Does the size distribution, from small to large, bear any relationship to the abundance distribution, from rare to common?

If you examine larger and larger areas, how will the number of plant species you see depend on the area you sample? Is the dependence of species' richness on area similar across all the summits and all the subalpine meadows?

The individuals within any species appear to be distributed across the alpine habitat in a somewhat non-random clumpy pattern. What mathematical distribution describes that clumpiness?

Is there some way one could estimate the abundances of the species throughout the entire alpine habitat of a mountain without having to look over the whole area and count all the little individual plants?

As you walk down the mountainside into the forest, you pass through open meadows in which the plants are generally larger and grow closer together; you wonder if the patterns there, the answers to the questions above, are similar to those in the alpine?

To answer these questions you need data. And to obtain data you need a strategy for obtaining those data, a strategy for censusing the plant communities. So when you return home at the end of that exciting day, you make a plan for censusing the mountain sides... a plan that you will carry out over the coming weeks with several assistants.

3.1.2 Censusing like a macroecologist

When you return to the mountain the next day you decide to begin censusing in the meadows within the forest. The reason is that spring comes earlier there than on the summits, and so the subalpine plants are more advanced phenologically; some are finished flowering but still recognizable, while others are only beginning to flower. By the time you finish the subalpine meadow censuses, more of the alpine plants will be in identifiable stages of their summer growth and you can census there.

You pick three subalpine meadows and in each mark out a randomly located 4 m by 4 m plot, gridded down to ¼ m by ¼ m cells. In each small cell, you census nearly all the plant species and their abundances; you can't measure the abundances of the graminoids because it is too difficult to decide what constitutes an individual plant, so you just note which grid cells the various species of graminoids occupy. Over the next couple of weeks, you complete those censuses and similar ones of several other mountains nearby.

The meadow censusing completed, you and your assistants now climb to the summit of that first mountain, and lay out on the south-facing slope, just below the summit, a 16 m by 16 m grid of square meter cells spanning a total area of 256 square meters. Because soils are thin, the growing season is short, and the wind can be ferocious, the plants here grow sparsely and so you can readily count the numbers of individual plants from each of plant species in each of the cells. Moreover, you estimate and record the approximate heights of a randomly selected subset of the individuals in each species.

In addition to the detailed plant census of everything growing in the 256 m^2 plot, you lay out an imaginary line, called a transect, from the summit plot down to where alpine habitat ends at tree line. Along the transect, at 100-m intervals, you set up a 2 m by 2 m plot called a quadrat, and then identify and list every plant growing in it. You do not have time to count the abundance of each species, so you will just collect presence–absence data. As you list the species, you note that there are definitely "hot spots" and "cold spots," meaning locations on the transect where many plants grow and other locations that are relatively barren.

Over the next few weeks you climb to nearby peaks to carry out similar alpine plant censuses, and again you list the species found in plots along transects down to tree line. Then, just to make sure you caught the species that do not begin their growth cycle until late in the summer, you return to a few of the sites and do some spot checks.

At the end of the summer you have several field books full of data, and with a GPS unit you obtain additional information, such as the elevations of, and the distances between, the different plots that you censused. You are well on your way to being a macroecologist! Even without the balloon ride, it's fun. In fact, you might at this stage think about spending the following summer in a completely different mountain range, perhaps on a different continent, so that you can determine the ubiquity of the patterns that you will soon discern from your data.

But now you have to think about how to use your data to answer the questions. You are going to have to search the data for patterns, and patterns suggest some sort of graphical representation. But what do you graph against what? This is what macroecological metrics are for, so let's discuss them.

3.2 Metrics for the macroecologist

By "metrics" I simply mean the functions that express relationships among types of data. For example, if you have two columns of data in your field book, one being the areas, A, of censused plots and the other being the number of species, S, observed in each of those plots, then the function $S(A)$ is a metric called the species–area relationship. If your field book contains data listing the abundances of each species in a collection of species, then if you rank the species from most to least abundant, that ordered list is a metric called the rank–abundance distribution.

Tables 3.1a and 3.1b introduce most of the macroecological metrics that we will address in this book. They show the symbols we use and provide capsule definitions of the metrics. The Glossary at the beginning of the book defines all the variables and symbols that we use throughout, including the independent variables that the metrics in Tables 3.1 depend upon. Following some general comments about units, notations, and forms of metrics, I take the reader systematically through the entries in Tables 3.1a and b twice. First, I carefully explain the meaning of each metric, and some relationships among them; if you are not comfortable with their meaning, the subsequent data graphs and theory will not make sense. On the second time through, Section 3.4, I describe the way or ways in which each metric is usually graphed and analyzed to extract information from data about pattern; and I describe in broad terms what the prevalent observed patterns actually look like. Finally, I discuss some of the applications of these metrics in conservation biology.

3.2.1 Units of analysis

While species are often the taxonomic unit of interest in macroecology, you need not confine your analysis of macroecological patterns to them. For example, you may find that a genera–area or family–area relationship is of more interest than a species–area relationship, or perhaps you may want to know the distribution of abundances over populations rather than over species. In the descriptions of the metrics provided here, where I write "species" you can substitute other taxonomic units, and where I write "individuals," you might think instead of colonies (e.g. for social insects) or some other aggregation of individual organisms.

Even after selecting species and individuals to be the units of macroecological analysis, you still have a further choice to make regarding which species you wish to include in your analysis. Do you include all the vascular plants in your censuses in the subalpine meadow plots, or do you restrict the dataset to just the non-woody forbs and grasses? Or perhaps you wish to focus on just lichens or mosses or beetles. The choices are many and yours should be guided by the scientific questions you are trying to answer. A comprehensive theory of macroecology should apply regardless of the choice of the system under consideration. The important thing is that you select an unambiguous category to study. For example, doing a census of seedling distributions in a forest might lead to difficulties deciding when a seedling becomes a sapling and so you would want to create an unambiguous criterion to decide what to count.

Table 3.1 Macroecological metrics. (a) Species-level metrics. (b) Community-level metrics. The variables upon which the listed distributions and dependence relationship are conditional (n_0, S_0, N_0, E_0, L_0, A_0) are explained in the text and in the Glossary. Chapter 7 will derive the actual dependence on conditional variables of the metrics.

Symbol and Name of Metric		Description of Metric
Table 3.1a		
$\Pi(n\|A, n_0, A_0)$	intra-specific spatial-abundance distribution	Probability that n individuals of a species are found in a cell of area A if it has n_0 individuals in A_0.
$B(A\|n_0, A_0)$	box-counting range-area relationship	Dependence on cell size of a box-counting measure of range for a species with n_0 individuals in A_0.
$C(A,D\|n_0, A_0)$	intra-specific commonality	Dependence on A and D of the fraction of pairs of cells of area A, separated by a distance D, that both contain a species with n_0 individuals in A_0.
$\Omega(D\|n_0, A_0)$	O-ring measure of aggregation	Average over each occurrence of an individual, of the density of individuals within a narrow ring at a radius D, divided by the density in the ring expected in a random distribution.
$\Theta(\epsilon\|n_0, S_0, N_0, E_0)$	intra-specific energy distribution	Probability density function for an individual from a species with n_0 individuals to have a metabolic energy rate between ϵ and $\epsilon + d\epsilon$.
$\Delta(D\|n_0, A_0)$	intra-specific dispersal distribution	Probability density function for an individual in a species with n_0 individuals in A_0 to have a dispersal distance between D and $D + dD$
Table 3.1b		
$\Phi(n\|S_0, N_0, A_0)$	species-abundance distribution (SAD)	Probability that in a community with S_0 species and N_0 individuals, a species has abundance n.
$\bar{S}(A\|S_0, N_0, A_0)$	species–area relationship (SAR)	Average number of species in a cell of area A if S_0 species in A_0
$\bar{S}(N\|S_0, N_0)$	collector's curve	Average number of species found in a random sample of N individuals.
$\bar{E}(A\|S_0, N_0, A_0)$	endemics-area relationship (EAR)	Average number of species unique to cell of area A if S_0 species in A_0.
$\bar{X}(A,D\|S_0, N_0, A_0)$	community-level commonality	Average fraction of the species in cells of area A that are found in common to two cells of area A a distance D apart, if S_0 species in A_0.
$\Psi(\epsilon\| S_0, N_0, E_0)$	community energy distribution	Probability density function for an individual in a community with $S0$ species, $N0$ individuals, and total metabolic rate, $E0$, to have metabolic rate between ϵ and $\epsilon + d\epsilon$.
$\bar{\epsilon}(n\| S_0, N_0, E_0)$	community energy-abundance relationship	Dependence of average metabolic rate of the individuals within a species on that species' abundance, n.
$\Lambda(l\|S_0, L_0)$	link distribution in a species network	probability that a species in a network with S_0 species and L_0 links is connected by l links to all other species

3.3 The meaning of the metrics

A quick look at Table 3.1 reveals that the metrics of macroecology divide into two categories: species-level and community-level. Species-level metrics describe various properties of single species within an ecosystem, while a community-level metric describes properties of a collection of species co-inhabiting the same ecosystem. The ecosystem referred to here could be anything from a small plot to an entire biome, such as the Amazon basin.

The metrics are of the form $f(X|Y)$, where X and Y may each refer to a single variable or to several variables; the notation is read as: "f of X, given Y". The notation is the standard one for a conditional function; it tells us that the dependence of the function f on the variable X depends on the values of the conditionality variables, Y. The choices of the conditionality variables in Table 3.1 are intended to be plausible but not established at this stage in the exposition. In Chapter 7, we shall see what METE actually predicts them to be. The quantities A_0, S_0, N_0, E_0, that appear as conditionality variables in some of the metrics in the Table, are what I introduced as state variables in Section 1.3.6.

By using this notation, I do not mean to suggest that in all applications, Y has been measured and the distribution of X is being predicted. In fact, we will see that there are instances in which the metric can be used to infer the value of Y from measurement of X.

There are two other notational matters to address. First, some symbols in Table 3.1 have a line above them, such as $\bar{S}$. I use this notation throughout the book to denote the average of a quantity. Second, when the conditionality of a metric is not essential for making a point, I will usually leave it out to avoid clutter.

Looking even more carefully at Table 3.1, you will note that some of the metrics describe probability distributions or probability density functions, and others describe dependence relationships. The distinction is important. A probability distribution is normalized to 1 and informs us of the relative likelihood of different values of its argument. A dependence relationship, in contrast, informs us of the relationship between two ecological variables, such as the dependence of the average metabolic rate of the individuals in a species on the abundance of the species. With one exception, I denote macroecological metrics that are probability functions with Greek letters and I denote the dependent and independent ecological variables in dependence relationships with Latin letters. The exception is the metabolic rate of an individual organism, which I denote by the Greek letter ϵ rather than by e to avoid confusion with the universal constant that is the base of natural logarithms. Finally, when I refer to generic probability distributions I will generally use Latin letters.

Dependency relationships can, and in several instances in the Table do, result from a probability distribution. For example, if the discrete probability distribution $p(z|x)$ is known, then the dependence on x of the mean of z (denoted by $\bar{z}$) over the distribution $p(z|x)$ is given by:

$$\bar{z}(x) = \sum_z z \cdot p(z|x). \tag{3.1}$$

$\bar{Z}(x)$ will be a function of x and is thus a dependence relationship. If the distribution p is continuous, then the summation in Eq. 3.1 is replaced by integration. Because of such connections between probability distributions and dependence relationships, the metrics in Table 3.1 are not all independent of one another. Nevertheless, I list them all separately here because each has been the focus of research interest.

Finally, note that the probability distributions fall into two classes according to whether the independent variable is discrete or continuous. A probability distribution, $p(n)$, where n is a discrete variable, such as species' abundance, and the sample space is a set of species, has the meaning that at each value of n, $p(n)$ is the probability that a random draw from the pool of species gives a species with abundance n. In this case, $p(n)$ could be estimated empirically as the fraction of the sample space possessing that discrete n-value. For a probability distribution defined over a continuous variable, $p(\epsilon)$, where ϵ is, say, metabolic energy, p is defined as follows: $p(\epsilon)d\epsilon$ is the probability that a random draw from the sample space has an energy value between ϵ and $\epsilon + d\epsilon$. A distribution over a continuous variable, such as p, is called a probability density function, which is sometimes abbreviated as a "pdf."

A practical problem can arise when comparing theoretical probability distributions with data. Consider a distribution over the integers, such as an abundance distribution. Empirically, the fraction of species with any particular value of n is likely to be zero for most values of n and so the empirical distribution will be a series of discrete spikes at particular integer values of n, separated by intervals of integers at which the value is zero. This complicates the task of comparing a relatively smooth theoretical distribution with real data. A similar problem arises with continuous data and probability density functions. To avoid this problem, when comparing theoretical with empirical distributions, it is often most insightful to examine something called a rank distribution that is constructed from the probability distribution. The means of doing that will be discussed in Section 3.4.

Now, let's look at the actual metrics.

3.3.1 Species-level spatial abundance distribution

Walking through a forest you cannot help but be struck by the fact that some trees grow close to others, while some are more isolated. At large scale you noted this tendency for inhomogeneity from the vantage of your balloon ascent. If you recognize the different species, you may observe that for any selected species there are patches with many individuals and other patches of the same size with with few or none. That could indicate that the individuals are randomly located, and by chance there are unpopulated patches, or it might indicate non-randomness. Our first species-level metric, $\Pi(n)$, informs us of the discrete distribution of the numbers of individuals across cells of area A for each species, and allows us to answer questions such as "is the pattern random?".

To ensure understanding, here is an example. Suppose that in one of those plots of area $A_0 = 256$ m^2 laid out just below the summit of a peak, we have a species with $n_0 = 340$ individuals. If we choose an arbitrary single cell or quadrat with area $A = 1$ m^2 from within the larger plot, what is the probability that in it will be $n = 12$ individuals of that species? The quantity $\Pi(12|1, 340, 256)$ answers that question. In Section 4.4.1 we will see how the shape of $\Pi(n)$ is a reflection of whether the individuals in a species are randomly distributed, or whether they cluster or repel one another. Exercise 3.11[CE1] gives you an opportunity to construct numerical values for this metric, for a made up landscape.

We always assume that $A < A_0$, but the prior knowledge of abundance might be at scale A rather than at scale A_0. For example, we might know from census data that at some scale A there is an average of n individuals per plot. What, then, is the distribution of abundances at some larger scale A_0? Box 3.1 shows how this is estimated.

Box 3.1 Inverting conditional distributions and upscaling spatial abundance distributions

Suppose we have a conditional discrete probability distribution $P(n|m)$, which is the probability of measuring n, given m. Similarly, we can express the probability of measuring m, given n, as $Q(m|n)$. If we know the form of one of these, what is the form of the other? First write a joint probability distribution $P(n,m)$. There are two equivalent ways to evaluate this:

$$P(n, m) = P(n|m)M(m) \qquad (3.2)$$

and

$$P(n, m) = Q(m|n)N(n). \qquad (3.3)$$

Here $M(m)$ and $N(n)$ are the unconditional distributions for obtaining values m and n, respectively. The functions M and N are not necessarily of the same form, which is why we have given them different names. We can express $N(n)$ as:

$$N(n) = \Sigma_m P(n|m)M(m) \qquad (3.4)$$

and hence

$$\begin{aligned} Q(m|n) &= P(n|m)M(m)/N(n) \\ &= P(n|m)M(m)/\Sigma_m P(n|m)M(m). \end{aligned} \qquad (3.5)$$

This is simply one way to express Bayes' law of conditional probabilities.

Box 3.1 (*Cont.*)

Returning to our spatial abundance distribution, and using a more explicit notation, we can now write the probability, $\Pi'(n_0|A_0, n, A)$, of there being n_0 individuals at some larger scale, A_0, given n individuals at a smaller scale, A, as:

$$\Pi'(n_0|A_0, n, A) = \frac{\Pi(n|A, n_0, A_0)M(n_0)}{\sum_{n_0=1}^{N_0} \Pi(n|A, n_0, A_0)M(n_0)}. \tag{3.6}$$

$M(n_0)$ would now be the unconditional probability that a species has n_0 individuals at scale A_0. Choosing that unconditional probability function can be difficult in some situations, yet Eq. 3.6 is useless without some choice for M. In our problem, we might take $M(n_0)$ to be the species–abundance distribution, Φ, at scale A_0 if we have a theory that predicts that metric. In Chapter 7 we show that MaxEnt theory predicts both Π and Φ, and hence Π'.

When I write the metric expressing the spatial distribution of individuals in a species in the form $\Pi(n|A, n_0, A_0)$ I am implicitly making a strong statement. Think of all the things that might influence the probability that there are n individuals in A. If it is a plant, perhaps the color of its flower, or the nature of its pollinator, or its dispersal mechanism, or the properties of the other species it shares space with, will influence that probability. Moreover, all those mechanisms listed in the first paragraph in Section 1.3.2 might matter. So I am not just defining metrics in Table 3.1, I am peeking behind the curtain to glimpse the theory that will be developed in Part IV. In that theory, only cell areas A, A_0, and the abundance of the species in A_0 influence the probability described by the metric. That is a huge simplification, so of course we will have to see if it is an unjustified over-simplification.

Available census datasets do not always include measured abundances; in many cases only the presence of a species is recorded in a given area. Take the typical case in which A is nested within A_0. Then, from the definition of $\Pi(n)$, we can express the probability of presence in A as $1 - \Pi(0)$. Equivalently, this expression is the expected fraction of cells of area A that are occupied by the species. This leads us to the second metric in the table, which relates to the range of a species, but it will be best understood if we first look generally at the problem of specifying a species' range and then glance at some ideas from the literature on fractals.

3.3.2 Range–area relationship

If you look at a Field Guide to the Birds of anywhere, you will likely find a range map for each species. The stippled or colored patch on the map tells you where, within the country covered by the guide, the species is found. But what is such a range map really telling you? The author likely outlined on a map of the country a free-form, but not too fragmented, or in other ways complicated, blob that both includes all the places where the species has been seen and minimizes to some extent inclusion of places where it has not been seen. So of course not every place within the stippled range is really suitable habitat. This field-guide range is sometimes called the geographic range of the species.

But you might seek a more detailed view of the bird's locations. For example, if you had census data on a checkerboard grid, fine enough to include, say, 10,000 cells falling within the country of interest, then a range map might show exactly which of all those cells had sightings. An even more detailed picture would result if you had data for 1,000,000 cells, however, and with such a fine grid you would get a different measure of the range.

The problem of specifying the range of a species poses the same ambiguities as does the problem of stating the length of a coastline. To explore the implications of that, I turn to fractals, not because the spatial distributions of individuals within species are actually fractals, but because the mathematics used to study fractals is powerful and applicable to our problem.

Start with line shapes rather than locations of objects on a two-dimensional surface. An irregularly-shaped line, such as a coastline, has a length that is ambiguous. In particular, suppose you measure the length, L, of the coastline the following reasonable way: you take a measuring rod of length l and you determine how many times, $N(l)$, you have to lay it down end over end to span the coastline. You then define length to be $L = N(l) \cdot l$. Reasonably, you ensure the rod is not so small that it probes spaces between small rocks along the shore, nor so large that you can't lay it out at least a few times along the coastline.

You will find that the value of L depends on the length of the measuring rod. This is because a smaller rod can go in an out of more little indentations in the shoreline, where a larger one would not "see" these indentations. Hence you expect that N decreases as l increases and, if it does not decrease exactly as l^{-1}, then L will depend on l. To proceed you make a graph of L versus l, or (for reasons explained in Box 3.3 in Section 3.4) of $\log(L)$ versus $\log(l)$. As discussed further in Section 3.4 and Appendix B, the shape of such graphs can inform us if the coastline is fractal in shape (Mandelbrot, 1982).

This description of the length of a coastline can be generalized to two dimensions to give a useful meaning of the range of a species, as measured from occupancy data. Suppose a North American bird species is located in patches throughout the continent. If the continent is covered by cells of area A, then in some fraction of those cells, the species will be found. So, instead of length l, think of cell area A, and instead of N being the number of times that you can lay down the stick, let it be the number of cells of area A that are occupied by the species. We can define a kind of "area of occupancy," $B(A)$, for the species by multiplying N, which will depend on A, by A: $B(A) = N(A) \cdot A$. Again, the shape of a $\log(B)$ versus $\log(A)$ graph informs us if the occupancy pattern is fractal. This measure of area of occupancy, B, is sometimes referred to as a "box-counting" measure. The reason is that if we think of each cell of area A as a box, then $N(A)$ is just a count of the number of occupied boxes. So our metric, B, describes the dependence of this measure of range on the size of the cells used to measure it (Kunin, 1998; Harte et al., 2001).

$B(A|n_0, A_0)$ can be obtained from the species-level abundance distribution metric $\Pi(n|A, n_0, A_0)$:

$$B(A|n_0, A_0) = [1 - \Pi(0|A, n_0, A_0)] \cdot (A_0/A) \cdot A. \qquad (3.7)$$

The first term on the right-hand side is the fraction of occupied cells of area A and the second is the total number of cells of area A within A_0. Their product is $N(A)$, the number of occupied cells. The third term is the area of such cells.

3.3.3 Species-level commonality

The third entry in Table 3.1a describes co-occurrences of individuals. Pick a species with n_0 individuals in A_0, and any two cells of area A that are a distance D apart. The fraction of such pairs of cells, in which both cells contain at least one individual from that species, is the metric $C(A, D|n_0, A_0)$. This metric describes a dependency relationship jointly on two variables. The notation anticipates that the value of this metric for any specified value of A and D will be conditional on the variable n_0 and A_0.

When $C(A,D)$ is viewed as a function of D, with cell area A held fixed, the metric is often referred to as a measure of "distance-decay." This nomenclature is appropriate because the function is often a decreasing function of inter-cell distance, reflecting the tendency for individuals in a species to be found close to, rather than far from, one another (Nekola and White, 1999).

There is an important distinction between metrics like commonality, C, and metrics like the species-level spatial abundance distribution, Π. The latter is a function describing information about a single cell, whereas the former describes what is happening in two cells at the same time. The former contains information about spatial correlation, while the latter does not.

There are actually many ways to construct generalizations of $\Pi(n)$ to provide more information than simply asking for the probability of a specified abundance level in a single cell. These can give us even more information than does $C(A,D)$. First, we might be interested in not just presence but actual abundances in the two cells: what is the probability that there are n_1 individuals in one of the cells and n_2 in the other? We might label that metric $C(n_1, n_2, A, D|n_0, A_0)$. Second, we might generalize this by asking for the probability of a specified set of simultaneous abundances in each of $M \geq 2$ cells that are of specified area and specified distances from each other. The maximum amount of information about spatial structure of the distribution of individuals in a species at a given scale of resolution (that is a given cell size, A, into which A_0 is gridded) is given by a function that describes the probability, $C(n_1, n_2, \ldots, n_K)$, of observing a set of assigned abundances $\{n_i\}$ in all of the cells. Because the plot can be gridded into A_0/A cells, the index i runs from 1 to $K = A_0/A$ and the spatial ordering of the cells, and thus the information needed to calculate the distances between them, will be reflected in their order listed in the argument of C. Predictions of the actual form of such a commonality metric can be obtained from macroecological theory only in very limited cases.

Related to the commonality metric $C(A,D|n_0, A_0)$ is a measure of how the density of individuals within a species changes with distance from any randomly selected individual in the species. To measure it, you pick an individual at random from the species and then look in the thin ring between two concentric circles, centered at the selected individual, and of radius r, and $r + \Delta r$. The number of individuals in that ring, divided by the area of the ring, is the density of individuals at a distance r from the selected individual. Dividing that density by the density that you would expect if the individuals in the species were randomly distributed throughout A_0, and then taking the average of that ratio for all individuals in the species gives the value of the metric. It is called an "O-ring" measure of aggregation. Condit et al. (2000) denote this metric by the letter Ω, or in our conditionality notation, $\Omega(D|n_0, A_0)$. A value greater than 1 at relatively small values of the radius r indicates that the conspecifics of each individual are more clustered in a small neighborhood surrounding each individual than is expected under random placement. A value less than 1 indicates that the individuals are over-dispersed locally, as expected, for example, for species that have evolved mechanisms for avoiding dense aggregates which might attract predators (Janzen, 1970; Connell, 1971).

Related to the O-ring metric is Ripley's K-statistic (Ripley, 1981). Whereas the O-ring measures density in an annulus surrounding an individual, Ripley's K measures density in the entire circle surrounding an individual. By considering an annulus as the region in between two concentric circles, and recalling that the area of a narrow annulus is $2\pi r \Delta r$, you can show (see Figure 3.1) that the density of individuals, $\rho_{\Delta r}(r)$, in an annulus of width Δr at radius r, is given by:

$$\rho_{\Delta r}(r) = \frac{\pi(r + \Delta r)^2 K(r + \Delta r) - \pi r^2 K(r)}{\pi(r + \Delta r)^2 - \pi r^2} \approx \frac{\frac{d[\pi r^2 K(r)]}{dr}}{2\pi r} \quad (3.8)$$

Because the O-ring metric is $\rho_{\Delta r}(r)/\rho_{\Delta r,\text{ random}}(r)$, the O-ring metric and the derivative of Ripley's K are related.

Finally, we note that instead of calculating the density of conspecifics in an annulus or a circle surrounding an individual, we could define a metric that describes the probability of a conspecific occurrence in the annulus or in the circle.

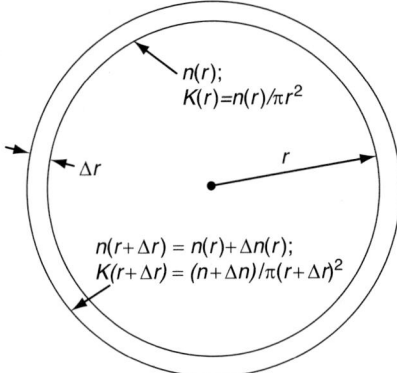

Figure 3.1 The space between these two concentric circles is the "O-ring" referred to in the text. $K(r)$ is the density of individuals in a circle of radius r, and the figure illustrates why Eq. 3.8 relating O-ring measure of aggregation to the derivative of Ripley's K statistic (circle density) holds.

3.3.4 Intra-specific energy distribution

You can look at a gathering of people and wonder which among them is using a lot of energy in their daily lives and which are using relatively little. You could go further and ask "What is the distribution of the rate of energy use across the group?" The same question can be asked of the individuals within any species in an ecosystem; $\Theta(\epsilon)$, a probability density function, is the answer. It describes the distribution of metabolic energy rates across the individuals of a species in an ecosystem. The species is assumed to have abundance n_0 in A_0 as with the other species-level metrics. Because it is an energy distribution over individuals rather than an abundance distribution over space, you might expect it will not depend on A_0 but might depend in some way on the total amount of energy utilized by the species in question, and perhaps on the total number of species and individuals in the system. In Table 3.1 this metric is shown as being conditional on total number of species, S_0, total abundance, N_0, and total metabolic rate, E_0; in Chapter 7 we will see that METE predicts that this is indeed the case.

Empirical determination of the independent variable in this distribution, the metabolic rate of an individual, requires some explanation. Measuring metabolic rates of animals is not easy, whereas measuring their body size or mass is relatively straightforward. Hence if one takes seriously a theory, such as the MST described in Section 1.3.7, that relates metabolic rate to body mass, then data on metabolic rate distributions can be inferred from mass distributions. Similarly, if theory predicts the form of the energy distribution Θ, a mass distribution can be derived if the MST result is assumed. Converting back and forth between mass and energy probability density functions requires care, however, so in Box 3.2 I show how this conversion is carried out.

Box 3.2 Converting probability densities to new independent variables

We are given a probability density function $f(x)$ and another variable, y, which can be expressed as a function of x: $y = y(x)$. Inverting $y = y(x)$, we can also write $x = x(y)$. For example, if $y(x) = x^2$, then $x(y) = y^{1/2}$. The probability distribution for y, $g(y)$, is given by:

$$g(y) = f\big(x(y)\big)(dx/dy). \tag{3.9}$$

The term dx/dy is needed to ensure that, if we integrate each distribution over equivalent ranges of their independent variable, we get the same result: $\int dx\, f(x) = \int dy\, g(y)$.

We apply this to $\Theta(\epsilon)$ to derive the mass distribution over individuals. Assuming a general scaling relation between mass and metabolism of the form:

$$\varepsilon(m) = c \cdot m^b, \tag{3.10}$$

we obtain the mass distribution, $\rho(m)$:

$$\rho(m|n_0,\ldots) = \Theta(c \cdot m^b | n_0,\ldots) \, b \cdot c \cdot m^{b-1}. \tag{3.11}$$

Just as $\Theta(\epsilon|n_0,\ldots)d\epsilon$ is the fraction of individuals in a species with abundance n_0 with metabolic energy rate in the interval $(\epsilon, \epsilon + d\epsilon)$, $\rho(m|n_0,\ldots)dm$ is the fraction of individuals with mass in the interval $(m, m + dm)$.

3.3.5 Dispersal distributions

Fledgling birds typically establish their nests some distance from the location of the nest where they were born. Seeds produced from a parent plant travel and germinate some distance from that parent. The concept of a dispersal distribution is often used slightly differently in plant and animal ecology. In plant ecology it usually refers to the distribution of distances that seeds travel from parent plant to final destination. However, most seeds never germinate and become viable plants. In animal ecology a dispersal distribution for, say, birds, is often taken to be the distribution of distances between the nest site where a bird is born and the site where the offspring eventually nests. That definition has a number of advantages in spatial macroecology, and so here I apply it to plants as well as animals. Thus the dispersal distribution for a species of plant is the distribution of distances between where seeds are produced and where the seeds produce germinants that grow to maturity. The two types of dispersal distribution can have quite different shapes, as, for example, would be the case if seeds that by chance land very far from a parent are less likely to be in a favorable habitat for germination and survival than they would be if they landed closer.

The way the conditionality of our dispersal metric, $\Delta(D|n_0, A_0)$, is written, it appears as if the distribution of dispersal distances is determined by the abundance,

n_0, of the species in A_0. In fact, a reverse interpretation generally makes more sense; the ultimate density of the species is likely to be determined by the distribution of dispersal distances. It is important to keep in mind that the form of the conditionalities in our metrics (i.e. which variables are to the left of the conditionality symbol "|" and which are to the right of it) should not be interpreted as informing us about the direction of influences. Referring back to the definition of the O-ring measure of spatial structure, however, it is plausible to assume that the slope of the dispersal distribution will influence the O-ring metric of aggregation.

In addition to there possibly being a relationship between abundance and the shape of the dispersal distribution, body size could also plausibly influence that distribution. In general, one would expect that large animals have the capacity to disperse longer distances than do smaller ones. Because abundance and body size are also related, as we will discuss in some detail later, and because a satisfactory theory of dispersal distributions is lacking, let's leave aside for now the question of what additional variables should appear to the right of the conditionality symbol in $\Delta(D|n_0, A_0)$.

3.3.6 The species–abundance distribution (SAD)

Ecosystems are fascinating in part because some species are very common and others quite rare. If you are a birder, you know both the thrill of spotting a rarity and the drama of, for example, a hundred thousand snow geese descending at dusk to their wintering grounds. Between commonness and rarity lies a probability distribution, and that distribution is among the most widely studied metrics in ecology. Reliable prediction of the shape of this metric, $\Phi(n)$, has been the goal of much macroecological theorizing. The SAD is a frequency distribution in the usual sense used in probability theory. Imagine a barrel filled with balls of different colors, where each color is a species. There might be just a very few yellow balls, but many red ones. If you reach in and pull out a ball, the chances are it will be of a color that is well-represented in the barrel. By repeatedly sampling the barrel, replacing the ball after sampling, and recording its color, you arrive at a probability distribution for the abundances. The data you obtain from a thorough and well-designed census, such as the one you carried out on the mountain-top plots, approximates spilling the contents of the barrel out on the ground and counting every ball that was in it.

3.3.7 Species–area relationship (SAR)

Like the SAD, this is a metric often studied in macroecology. A particular widely assumed functional form for the shape of the SAR, discussed later in the chapter, has been called by some a law of ecology. The realization that species' richness increases with sample size probably dates back millennia, but measurement of an SAR, along with an explicit proposal for its mathematical form, was first published by Olaf Arhennius (1921).[1] For a well-written review of early observations and theory pertaining to the SAR, see Rosenzweig (1995).

The idea behind the SAR is simple. As you look at larger and larger areas of habitat, with each larger area including the smaller areas, you generally find more species. When a census is carried out in that way, the SAR obtained is called a nested census, because small plots are nested within large plots. That species' richness increases with area for a nested census does not sound like a big deal, for unless the individuals in every species were found everywhere throughout the largest area studied, which we know is unlikely, it is obvious that larger areas will contain more species. But there is more to the SAR than that.

First, different reasons might be invoked to explain the increase in species' richness with census area. Larger areas generally contain more types of environmental conditions, such as values of soil moisture or slope or aspect, and this greater number of abiotic niches may support more species. Moreover, more migrants and dispersers from elsewhere are more likely to arrive in a larger area, and thus by chance alone there could be more species in a larger area. Because more individuals of a given species can be supported in a larger area, and because local extinction is more likely when population size drops below some critical threshold, extinction rates may also be lower in a larger area. Finally, there might be some inherent role played by area that results in more species able to co-exist within a larger area even when the habitat does not vary within it and the other mechanisms listed above are not operating.

Second, the actual shape of the rising SAR might inform us about a number of ecological phenomena: inter- and intra-specific competition, the distances different plant species' seeds are dispersed or animals migrate, the nature of the mosaic of soil types or land-surface topography or other abiotic landscape features, and more generally the forces that structure the partitioning of resources among species.

Third, the actual shape of the SAR has enormous implications in conservation biology. I discuss some of them in Section 3.5 and return to that topic again in Chapter 9.

Fourth, there are other types of SARs, in addition to nested ones. Censusing islands provides one frequently encountered way to examine non-nested SARs. One might expect that if species' richness is compared on islands of different sizes, but all approximately the same distance from a mainland, and if the habitats are not very different on the islands, then a large island is likely to, but will not necessarily, contain more species than a small one. The shapes of non-nested and nested SARs can differ, and quite generally we expect the shape of the SAR to differ depending on the census design. A good review of different types of SARs and the need to avoid conflating them, is found in Sandel and Smith (2009) and in Dengler (2009).

---

[1] He was the son of the Nobel Laureate Svante Arhennius, who among many other achievements carried out, in 1896, the first reliable calculation of the effect of increasing atmospheric carbon dioxide on Earth's temperature (Arhennius, 1896)...a remarkable family, indeed!

Consider the island SAR mentioned above. Larger islands generally contain more individuals and more species than smaller ones, at least if other factors, such as habitat quality, and distance to a mainland that acts as a source of organisms that disperse to the islands, are the same. The island SAR describes the dependence on island area of the number of species on each island. Area may or may not explain most of the inter-island variation in species' richness, depending on what other factors, such as those above, are at work. These considerations were the basis for the theory of island biogeography first developed in the 1960s (MacArthur and Wilson, 1967). We note that the islands need not be actual islands; they could be geographically non-overlapping patches of different areas marked out within a mainland ecosystem. The important thing about these real-island or conceptual-island SARs is that the census design involves independent sampling at each area, in contrast to the nested SAR census design.

Next, consider the notion of a collector's curve, $\bar{S}(N)$. This is not strictly speaking an SAR because its operational definition does not depend on area. The collector's curve describes the dependence on N of the number of species found in a random sample of N individuals. The shape of the collector's curve is determined by the distribution, $\Phi(n)$, of abundances in the community of species. The idea is simple. If you have a collection of individuals from a variety of species, say on a mainland, you can imagine putting them all in a barrel and then drawing individuals from the barrel, with or without replacement. As you draw more and more individuals, you find more and more species. The species list will not grow as fast as the number of sampled individuals of course, because there will be many repeats of the same species. As the list of species grows, you can graph the number of species versus the number of individuals collected and the shape of the graph is called a collector's curve or an accumulation curve. In the field, a collector's curve can be obtained either with or without replacement of individuals. Thus if you are trapping insects and killing them in the process, the collector's curve you obtain would be without replacement. On the other hand, if you simply observe without disturbing animals one at a time and record the growing list of species, you would obtain a collector's curve with replacement.

There is a connection between the collector's curve and the island SAR: the islands can be considered to be collectors and the mainland to be the barrel of individuals. The island collects individuals as they migrate or disperse from the mainland to the island and settle there. Suppose the number of individuals on an island is proportional to island area, and also suppose that individuals from each species migrate from the mainland to the islands in proportion to their numbers on the mainland. Then the contents of the mainland barrel uniquely determines the island SAR; the island SAR is just a collector's curve and is entirely determined from the shape of the abundance distribution $\Phi(n)$. If patches of varying area are marked out randomly on the mainland, the same idea could apply; the species' richness in each patch could be described by a collector's curve resulting from the individuals that happen to collect in those patches from all the surrounding habitat. These ideas are illustrated in Figure 3.2.

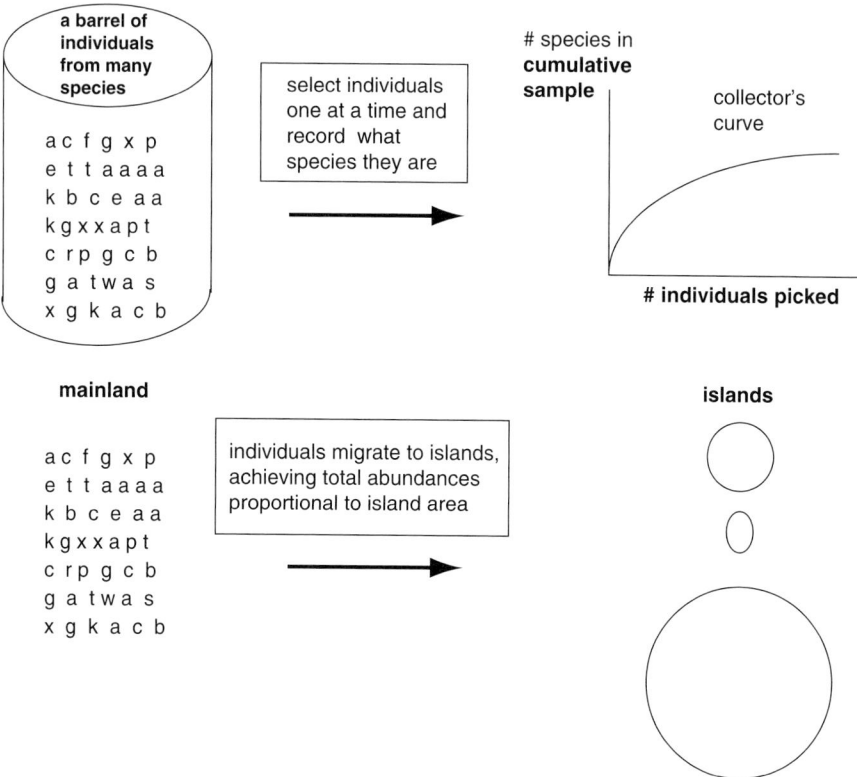

Figure 3.2 The concept of the collector's curve and its possible relevance to island species–area relationships.

The nested SAR is sometimes referred to as a mainland SAR, but of course one could conduct a nested census on either an island or the mainland. In a complete nested design, a large plot is typically laid out and either gridded into small cells that cover the entire plot area or each individual is point-located within the larger plot. In that latter case, a conceptual grid can be laid upon the plot and each point-located individual placed in a grid cell. The numbers of species in each of the small grid cells is recorded, and then the average of these numbers over all the small grid cells is taken to be the number of species at the scale of the small grid cell. Then the number of species in aggregates of grid cells are recorded, and averaged to give species' richness at that aggregate scale. Repeating this procedure for aggregates of many different sizes gives a nested SAR for the plot.

There is a fundamental reason why the collector's curve is not likely to describe the nested SAR, even while it may describe island SARs. The reason is that the distribution in space of the individuals in each species will affect the shape of the SAR. In fact, an expression we can write relating the nested SAR to two other metrics we have defined makes explicit the role that spatial structure and abundance distributions play in determining the shape of the nested SAR.

To motivate the expression consider an average cell of area A within some larger plot. The probability that a species with n_0 individuals in the larger plot will be found in the cell is 1 minus the probability that it is absent from the cell, or $[1 - \Pi(0|A, n_0, A_0)]$ (see Section 3.3.1). So if we assume that the species-level spatial abundance distributions are independent of one another across the species, then we can write:

$$\bar{S}(A|N_0, S_0, A_0) = \sum_{species} [1 - \Pi(0|A, n_0, A_0)] \quad (3.12)$$

Eq. 3.12 is simply summing the probabilities of occurrences of each of the species to give the expected number of species. This is the formula that you would use to derive $\bar{S}$ if you knew the actual values of the abundances of the species; the sum over species is actually a sum of $1 - \Pi(0)$ over the known values of abundance.

If you do not know the actual values of the abundances but you do think you know the abundance distribution, $\Phi(n)$, then Eq. 3.12 is replaced by:

$$\bar{S}(A|N_0, S_0, A_0) = S_0 \sum_{n_0=1}^{N_0} [1 - \Pi(0|A, n_0, A_0)] \Phi(n_0|N_0, S_0, A_0) \quad (3.13)$$

In this expression, we are summing over n_0 the product:
[the probability that a species with n_0 individuals in A_0 is present in A] × [the fraction of species in A_0 with abundance n_0].

This gives the expected fraction of species found in A. Multiplying that fraction by the total number of species in A_0 gives us $\bar{S}(A)$, the expected number of species in area A. Note that the upper limit of the summation in Eq. 3.13 is N_0. Strictly speaking it should be $N_0 - S_0 + 1$, because every species has to have at least 1 individual and so no species can have more than $N_0 - S_0 + 1$ individuals. However, in practice, this will not make a difference because N_0 is generally $\gg S_0$, and we will see that $\Phi(n_0)$ always falls to zero very rapidly at large n_0.

With Eq. 3.13 in hand, we can now also write an expression for the collector's curve. A barrel of individuals, from which one at a time is randomly drawn, has no spatial structure. Hence, if the spatial distribution $\Pi(n)$ in Eq. 3.13 is taken to be a binomial distribution for each species, thus implying a random draw of individuals, we should obtain the collector's curve.[2] In Section 7.8, we will show how to calculate collector's curves that way.

---

[2] Readers unfamiliar with the mathematical forms and properties of the major probability distributions, such as the Gaussian, the Poisson, the binomial, and the lognormal, can find a good summary in Evans et al. (1993); Jaynes (2003) is an excellent text on probability theory.

3.3.8 The endemics–area relationship (EAR)

A species is said to be endemic to a region if it is only found there. Thus we say that about 50% of the species of birds in Madagascar are endemic to that nation—you will not find them elsewhere in the world. The way I am going to use the term in this book is related but slightly different. Consider a plot such as the 256 m^2 plot on one of those mountain peaks. If we look at a cell of area 1 m^2 within that plot, we may find that it contains one or more species that are only found in that one cell and nowhere else in the remaining 255 m^2. Those species within the plot that are found only in that single cell we will call "endemic" to that single cell. Those species may well be found outside our 256 m^2 cell, but that is irrelevant to us here. Now we can do the same thing for all the cells of area 1 m^2 and then average the number of such endemic species over all 256 of the cells. That average value goes into the construction of the endemics–area relationship... it gives us $\bar{E}(1\ \text{m}^2)$. If we do this for cells of all areas within the plot, we have $E(A)$, the endemics–area relationship. Note that $\bar{E}(A_0) = S_0$.

Analogously to Eq. 3.13, we can write an expression for $\bar{E}(A)$ in terms of the species-level spatial abundance distributions and the species-abundance distribution.

$$\bar{E}(A|N_0, S_0, A_0) = S_0 \sum_{n_0=1}^{N_0} [\Pi(n_0|A, n_0, A_0)] \Phi(n_0|N_0, S_0, A_0) \quad (3.14)$$

The rationale for this formula is similar to that for Eq. 3.13, except that now the species-level spatial abundance distribution is evaluated at $n = n_0$. To see why, recall that a species with n_0 individuals in A_0 is endemic to a cell of area A if all n_0 of its individuals are found in that cell. The probability of that event occurring is exactly $\Pi(n_0|A, n_0, A_0)$.

3.3.9 Community commonality

This metric describes the co-occurrence of species. Consider all possible pairs of cells of area A that are a distance D apart. For any such pair determine the number of species found in common to the two cells and then average that number over all the pairs. The metric $\bar{X}(A, D|S_0, N_0, A_0)$ is that average divided by the average number of species found in a cell of area A. Paralleling Eq. 3.13 and 3.14, it can be related to a sum over abundances of the species-level commonality metric:

$$\bar{X}(A, D|S_0, A_0, ...) = \frac{S_0 \sum_{n_0=1}^{N_0} C(A, D|n_0, A_0, ...) \Phi(n_0|N_0, S_0, A_0, ...)}{S_0 \sum_{n_0=1}^{N_0} [1 - \Pi(0|A, n_0, A_0, ...)] \Phi(n_0|N_0, S_0, A_0, ...)} \quad (3.15)$$

Referring to Eq. 3.13, the denominator on the right-hand side of Eq. 3.15 is the average number of species that are found in a cell of area A. Applying the same reasoning that led us to Eq. 3.13, and recalling the definition of the species-level commonality, $C(A,D)$, we see that the numerator is the number of species that are found in both cells of area A. The metric $\bar{X}$ is called the Sorensen index.

Related to our metric of community-level species' commonality is the concept of beta diversity, a notion intended to describe species' turnover. One way to quantify turnover is with a metric $\bar{T} = 1 - \bar{X}$. Like $\bar{X}$, $\bar{T}$ can range from 0 to 1, with a value of 1 signifying that two patches have no species in common, so that turnover is complete. The term beta diversity is often contrasted with alpha and gamma diversity. Alpha diversity refers to local diversity, or average species' richness in, typically, a small patch; gamma diversity is the variety of species found across multiple habitats, so it is thought of as a large-scale measure. In that context, beta diversity is diversity still within a habitat type but over some distance. Because of the difficulty quantifying habitat types, and the likelihood that different species perceive habitat transitions differently, it is difficult to decide whether a change in slope or aspect, say, within a large forest warrants calling the diversity across that change beta or gamma diversity. We prefer to stick with our metric $1 - \bar{X}$ as a measure of changing diversity across space that can be applied at any scale.

The species-level O-ring metric described in Section 3.3.3 can also be generalized to community level. For this purpose consider a small cell of area A, and symmetrically surrounding it a thin annulus between radii r and $r+\Delta r$. The fraction of the species in the cell of area A that are also found in the annulus is yet another measure of community-level commonality as a function of distance.

3.3.10 Community energy distribution

$\Psi(\epsilon|S_0, N_0, E_0)$ is the community-level generalization of the species-level energy distribution $\Theta(\epsilon)$. In particular, Ψ is the probability density function for the distribution of metabolic energy rates over all the individuals in the community. Just as with Θ, testing of a predicted form for Ψ requires either difficult measurement of actual metabolic rates or straightforward measurement of body masses combined with theory, such as the MST, that relates body mass and metabolism. Ψ can also be converted to a distribution of the masses of individuals in the community using the MST, by the method shown in Box 3.2. Using $\epsilon = cm^b$, we get:

$$\Xi(m|S_0, N_0, E_0) = \Psi(c \cdot m^b|S_0, N_0, E_0) \cdot b \cdot c \cdot m^{b-1}. \tag{3.16}$$

3.3.11 Energy– and mass–abundance relationships

Imagine you are walking through a forest noting the numbers of trees in different size classes. First, suppose you pick some species that you can readily identify, even as a seedling. You will probably find many small seedlings of that species, fewer saplings, still fewer medium-sized trees, and interspersed among all these plants an even smaller number of fully grown trees, near the end of their lifespan. You would expect that, just on the basis of demographics. At every age, death can occur, and so, if the forest is in demographic steady state, the numbers must decline with age, and therefore, on average, with size. If the individuals are binned in equal size-intervals, a bar graph of numbers of individuals versus size would show bar height rapidly declining with increasing size; if logarithmic size intervals are used, the decline would be much less and conceivably the graph would appear flat or even show bar height rising with size.

There is a second way in which you could examine the distribution of sizes of trees. You conduct the same survey but you don't distinguish among species. Now all trees of each size class are counted. Would you expect to see the same size distribution that you witnessed for the individuals in a single species? A reason why you might not is apparent. Some trees never attain great size, becoming full grown at the same size as an adolescent individual of a tree that will grow huge. So, if those trees destined never to grow large are particularly abundant, then that will skew the distribution toward even more rapid decline in numbers with size.

As a third choice, you could examine the relationship between the numbers of trees of each species and the average mass of the individuals in each species. A priori there is no reason to think that that relationship would resemble either the first or second relationship above.

Each of these relationships can be considered to be a mass–abundance relationship (MAR) and they are all of interest. You could of course carry out the same analyses with animal data, whether for fish, birds, mammals, or any other broad category of life. In those cases you might not expect to see the same patterns as you do with trees for two reasons: trees grow until they die, while animals generally do not, and the ratio of the sizes of full-grown large trees to seedlings is much greater than the ratio of the size of mature mammals or birds to the size of their new-born.

Our metrics allow us to examine these various relationships more quantitatively and systematically. To each species we can assign a measured value of the average metabolic rate of its individuals, $\bar{\epsilon}$, and the abundance of that species, n_0, in A_0. The relationship between these variables is the metric $\bar{\epsilon}(n_0 | \ldots)$. Using Eq. 3.1, this metric can be derived from the intra-specific energy distribution, $\Theta(\epsilon | n_0)$. For convenience we leave out the subscript on n:

$$\bar{\epsilon}(n) = \int d\varepsilon \, \varepsilon \, \Theta(\varepsilon | n). \tag{3.17}$$

We can also take the intra-specific mass distribution $\rho(m)$ derived in Box 3.2, Eq. 3.11, from $\Theta(\epsilon)$, to convert to a relationship between the average mass of the individuals in a species, $\bar{m}(n)$, and the abundance of that species:

$$\bar{m}(n) = \int \mathrm{d}m\, m\, \rho(m|n). \tag{3.18}$$

From these energy–abundance and mass–abundance relationships, probability distributions for energy and mass, defined over the sample space of species, can be derived from the abundance distribution.

For example, starting with the species abundance distribution, $\Phi(n)$, and Eq. 3.9, the distribution over the species of the average metabolic rates of the individuals within species (which we denote by $v(\bar{\epsilon})$) is given by:

$$v(\bar{\epsilon}) = \Phi(n(\bar{\epsilon}))(\mathrm{d}n/\mathrm{d}\bar{\epsilon}). \tag{3.19}$$

The term $(\mathrm{d}n/\mathrm{d}\bar{\epsilon})$ in this expression is obtained by inverting the function $\bar{\epsilon}(n)$ in Eq. 3.17 and differentiating. In some cases this inversion procedure can be carried out analytically, while in other cases it has to be done numerically. It is important to keep in mind that $v(\bar{\epsilon})$ will be properly normalized if the values of $\bar{\epsilon}$ are allowed to range between $\bar{\epsilon}(n_{\max})$ and $\bar{\epsilon}(n_{\min})$, where $n_{\max}$ and $n_{\min}$ are the limits of the range of n used in normalizing $\Phi(n)$.

Paralleling Eq. 3.19, from Box 3.2 and the SAD, $\Phi(n)$, we can write an expression for the distribution, over the species, of the average masses of the individuals in species, $\mu(\bar{m})$:

$$\mu(\bar{m}) = \Phi(n(\bar{m}))(\mathrm{d}n/\mathrm{d}\bar{m}), \tag{3.20}$$

where $n(\bar{m})$ is obtained by inverting the function $\bar{m}(n)$ in Eq. 3.18.

Because of the proliferation to the edge of confusion of notations for mass and energy metrics, and the potential for confusing distributions with dependence relationships, I summarize all the notations in Table 3.2 and provide the location in the text that defines the metric, or relates the mass to the energy metric, or relates the dependence relationship to the probability distribution, in each case.

The sizes or masses of organisms are more readily measured than their metabolic rates, so we focus the rest of this discussion around mass–abundance relationships. Equation 3.18 provides one such relationship, where $\bar{m}(n)$ is the relationship between the abundance of a species and the average mass of its individuals. But there are actually several types of size- or mass–abundance relationships (MAR). A careful effort to distinguish and define them can be found in White et al. (2007). Table 3.3 summarizes these different types, following the terminology introduced by White et al., except that we replace their term "size" with "mass."

The term "density," which is abundance per unit area rather than abundance itself, appears in the definitions of the LMDR and the GMDR (in Table 3.3). The significance of this distinction helps us understand the difference between the local and global distributions. In particular, the empirical GMDR is obtained by examining species' distributions over continental or global scales and, for each species, estimating its average density only from locations where it is found. In fact, often the

Table 3.2 Summary of metrics describing metabolic rates and masses.

Level and meaning of distribution or dependence relationship	Energy		Mass	
Intra-specific: distribution of energy or mass across individuals within species	$\Theta(\epsilon)$	(Table 3.1a)	$\rho(m)$	(Eq. 3.11)
Inter-specific: distribution of energy or mass across all individuals	$\Psi(\epsilon)$	(Table 3.1b)	$\Xi(m)$	(Eq. 3.16)
Inter-specific: dependence on abundance of energy or mass averages over individuals within species	$\bar{\epsilon}(n)$	(Eq. 3.17)	$\bar{m}(n)$	(Eq. 3.18)
Inter-specific: distribution of energy or mass averages over individuals within species	$\nu(\bar{\epsilon})$	(Eq. 3.19)	$\mu(\bar{m})$	(Eq. 3.20)

Table 3.3 Types of mass–abundance relationships.

MAR	Definition	Relation to our Metrics
Individual mass distribution (IMD)	The probability distribution of masses of individuals in a community ... i.e. fraction of all individuals in a mass interval.	$\Xi(m)$ from Eq. 3.16.
Local mass–density relationship (LMDR)	Relationship between average mass of the individuals in a species and its population in a specified ecosystem.	$\bar{m}(n)$ in Eq. 3.18 inverted to give population size versus species mass: $n(\bar{m})$, with all populations in same community
Cross-community scaling relationship (CCSR)	Relationship between average mass of an individual in some collection of species and the total number of individuals.	N_0 dependence on average mass of individual in a community $= \int dm \cdot m \cdot \Xi(m)$
Global mass–density relationship (GMDR)	Relationship, across a larger community (perhaps of continental size) between average mass of the individuals in a species and its average population density.	$\bar{m}(n)$ in Eq. 3.18 inverted to give population size versus species mass: $n(\bar{m})$, with populations taken from anywhere on Earth.

locations where a species' density is estimated are locations where that species is relatively abundant, and so the density used to evaluate the GMDR is closer to maximum than average density. In contrast, the empirical LMDR is obtained by circumscribing a region and then within it estimating the density = (abundance within region)/(area of region). The same area is used for all species being compared, and so density is just proportional to abundance within the region.

Let us now relate the MARs in Table 3.3 to the metrics we have already defined. Rather than assume a specific scaling exponent, such as ¾, for the relationship between metabolic rate and body mass, we will assume a general form $\epsilon = c \cdot m^b$. An even more general relationship, not of power-law form, could also be used of course.

The IMD is given by Eq. 3.16: $\Xi(m) = \Psi(c \cdot m^b) \cdot b \cdot c \cdot m^{b-1}$.

The LMDR is given by $\bar{m}(n)$ in Eq. 3.18, where $\rho(m)$ in that equation is given in terms of our metric $\Theta(\epsilon)$ by Eq. 3.11.

The CCSR is obtained by determining the dependence on total abundance, N_0, of the average mass of the individuals in the community: $\int dm \cdot m \cdot \Xi(m)$. The distribution $\Xi(m)$, given by Eq. 3.16 in terms of $\Psi(\epsilon)$, will depend on N_0 as a consequence of the conditional dependence of the metric $\Psi(\epsilon)$ on N_0. The form of that dependence will, as we shall see, emerge from theory.

The GMDR is, in principle, related to our metrics the same way the LMDR is, but with data obtained at larger scales. However, in practice, it is neither easily related to our metrics nor very systematically measured. The reason is that for this MAR, the densities of species are compiled from a variety of censuses carried out in different locations, with density determined within areas of differing magnitude, and with different criteria used to select the density actually plotted against mass. A more systematic but more difficult way to look at a large-scale mass–density relationship would be to adopt the same approach as is used in determining empirical LMDRs, but execute it at much larger geograhic scale. Methods of estimating abundance at large scales, such as those I will describe in Chapter 7, could be used to do this. As White et al. (2007) emphasize, however, this would result in a measure of density that differs from that used in current data compilations to determine the GMDR. For reasons discussed by White et al. (2007), and also in Section 3.4.10, the GMDR poses an interesting conundrum for ecologists.

3.3.12 Link distribution in a species network

Many kinds of species' networks can be envisioned. The most frequently studied are food webs, in which the species in an ecosystem are the nodes in the network and the links between nodes are trophic interactions. If species A eats species B, then a link exists between node A and node B. A directed link has an arrow on it, indicating the direction of energy transfer. A web or network is called "quantitative" if the links are labeled with the actual magnitudes of energy flow between nodes. In the language of state variables, a food web can be minimally characterized by the total number of nodes, S_0, and the total number of links, L_0. If we knew more about the structure of the web, such as the number of herbivores, carnivores, omnivores, those numbers might also be considered as state variables.

Another kind of ecosystem web, classified as a bipartite network, describes the interaction between two groups of species. Consider pollinators and plants. Each plant species in an ecosystem is linked to some number of pollinator species and each pollinator species is linked to some number of plant species. Two link distributions can now be envisioned: the distribution of pollinator links to plants and of plant links to pollinators. Appropriate state variables for this system could be the number of links, the number of plant species, and the number of pollinator species.

In any network, some nodes may be connected to very few other nodes, while others may be highly connected. In an ecosystem, generalists feed on many species, while specialists feed on few. Thus a distribution of link numbers can be defined, and for a food web that is what $\Lambda(l|S_0, L_0)$ measures. $\Lambda(l|S_0, L_0)$ describes the distribution of the number of linkages, out of the total of L_0, connected to each of the S_0 nodes. These linkages could either be incoming (from a food source for the node), or outgoing (to a consumer of the node), or both.

3.3.13 Two other metrics: The inter-specific dispersal–abundance relationship and the metabolic scaling rule

We can define a function $\bar{D}(n)$ describing the dependence of the average dispersal distance of the individuals within a species, $\bar{D}$, upon the abundance, n, of that species. This metric can be obtained from the intra-specific dispersal distance distribution $\Lambda(D|n)$ by applying Eq. 3.1:

$$\bar{D}(n) = \int dD \cdot D \, \Lambda(D|n). \tag{3.21}$$

In addition, from $\bar{D}(n)$ and the SAD (Section 3.3.6), a probability distribution, $\eta(\bar{D})$ for the average dispersal distances across species can be obtained:

$$\eta(\bar{D}) = \Phi(n(\bar{D}))(dn/d\bar{D}), \tag{3.22}$$

where $n(\bar{D})$, the argument of Φ, is obtained by inverting the expression $\bar{D}(n)$ in Eq. 3.21

The dependence of metabolic rate on body mass, $\epsilon(m)$, is also not in Table 3.1. This metric does not fall into either the species-level or community-level category, but rather can be thought of as an individual-level metric, depending on factors such as individual physiology. In Section 1.3.7 we discussed the power-law behavior for this metric, as advanced by Metabolic Scaling Theory.

3.4 Graphs and patterns

Now I summarize prevailing patterns exhibited by the metrics we have just discussed. The literature describing patterns in macroecology is too voluminous to cover in anywhere near its entirety here; two recommended books on the topic are Brown (1995) and Gaston and Blackburn (2000).

Our knowledge of these patterns is based on many datasets describing a wide range of taxonomic groups, habitats, and spatial scales. Access to one such vegetation dataset from a serpentine grassland is provided in Appendix A, and you will become familiar with these data because several of the homework exercises I provide require their use. The data are available for your own use, but an explicit acknowledgment (described in the Appendix) is required for publication in any form of results obtained from them.

In this section I focus on generic patterns, and use specific census results, such as the serpentine data, only as examples. In Chapter 8, where empirical tests of METE are presented, I use a wide variety of actual datasets, including census data from different habitats, different taxonomic groups, and very different spatial scales.

Before delving into the shapes of metrics, two important mathematical issues need to be addressed. The first involves the use of log–log plots to exhibit mathematical patterns. As mentioned in Section 3.3.2, range–area relationships, and as we shall see, species–area relationships, are often plotted on log–log axes. The reason for doing so arises because we often seek to test for power-law behavior. Box 3.3

Box 3.3 Power-law determination

Consider the function $y = ax^b$, which is a generic version of proposed power-law models for the various phenomena listed in Table 1.1. If b is not equal to 1, a graph of this function, with y on the vertical axis and x on the horizontal axis, will be a curve; by staring at the curve it will be hard to determine whether the function really is described by a power-law function and, if so, what the value of b is. The same applies to real data as well as to pure functions; given data for y and x, you would like to know how well they follow a power-law relationship. You could perform a non-linear regression but a more commonly used and convenient method is to graph $\log(y)$ against $\log(x)$. Here's why.

Taking the logarithm of both sides of $y = ax^b$, we get $\log(y) = \log(a) + b \cdot \log(x)$. This is true regardless of the base of the logarithm. In this book, we will generally use logarithms to the base e, otherwise known as "natural logarithms." Often the natural log of x is denoted $\ln(x)$, but throughout this book I use the notation $\log(x)$ to refer to the natural logarithm of x. If I need to use some other base, such as 10, then I will write the base explicitly as $\log_{10}(x)$.

If the data are plotted as $\log(y)$ versus $\log(x)$, then a direct test of the power-law model is obtained with a linear regression. The slope parameter determined from the linear regression is a fitted value of the exponent b, and the regression coefficient R^2 equals the fraction of the variance in $\log(y)$ that is explained by the variance in $\log(x)$; it is widely interpreted as a standard measure of the goodness of the fit to a power-law function.

Caution is needed, however. Consider the log–log graph of the nested species–area relationship data in Figure 3.3. Ecologists are not very used to seeing 2-parameter models fit data with an R^2 as large as 0.9899, even with only 9 data points. If all you were told about the power-law fit was the value of R^2, you might leap to the conclusion that the data obey power-law behavior. Because fractional power-law behavior is closely related to the notion of self similarity and fractals (Mandelbrot, 1982; Ostling et al, 2003; Appendix B), you might further conclude that the system was fractal in its underlying geometry. That in turn might lead you to propose that some mechanism known to generate fractal patterns is operating in the system you are studying.

Box 3.3 (*Cont.*)

Look again at the graph. First, the data do not span a very large range of variability in the y-axis variable; the log of $\bar{S}$ has a range of only 1, meaning that there is only an e-fold range of variability in that variable. Without more variability in both the x and y variables, some would argue that straight-line fits are too easy to obtain. If there are very few data points spanning the range of the data, then it is easy to get a value very near 1 for the regression coefficient. Indeed, with only two data points, R^2 is identically 1. But regardless of the range of the data, as the number of data points increases, then the regression coefficient is increasingly informative. So the spanning range of the data may not necessarily inhibit statistical confidence in the power-law model. But it does have some bearing on the generality of the claimed power-law fit. It is just not very interesting to know that a species-area curve has a power-law fit over just a small range of variation in area or species' richness.

A more important reason for caution about over-claiming the goodness of power-law fits has to do with the actual curvature that can be masked by a near-1 value of R^2. Looking again at Figure 3.3, there is clearly systematic negative curvature. The slope of the line connecting the last two data points on the graph is 0.16, while the slope connecting the first two is 0.29. That difference in slopes, combined with the systematic trend toward decreasing slope as A increases, should shout caution to you about adopting a power-law model and claiming fractality. And if there were 10 times more data points filling in the gaps between the data points on the graph, the slope variation would still exist, even though to the casual eye the straight line fit would still look impressive.

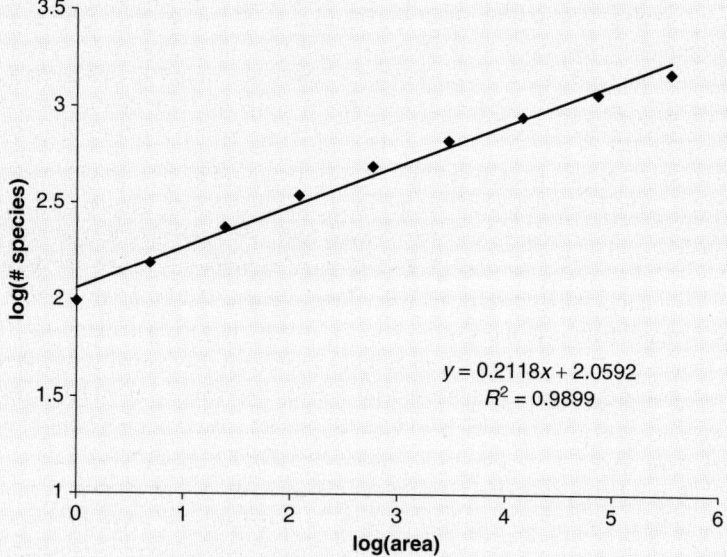

Figure 3.3 The R^2 value for a straight line fit to this hypothetical SAR is 0.9899, which is quite close to 1. While this suggests power-law behavior, distinct curvature is apparent. The slope between the first two data points is 0.29, while the slope between the last two is 0.16.

Box 3.3 *(Cont.)*

Had the residuals around the straight-line power-law fit to the data been randomly distributed, rather than falling below the line at the end points and above the line in the middle (indicating systematic curvature), you would be more justified arguing that a power-law model was appropriate. I emphasize this point because claims for power-law fits to species–area and range–area relationships abound in the literature. This would be harmless were it not for the consequence that once the power-law model begins to possess the force of law, it results in extremely misleading extrapolations to areas larger than can be actually censused. And this can result in misleading conclusions in conservation biology, as discussed in Section 3.5. Clauset et al. (2009b) provide a useful analysis of problems that arise in making claims of power-law behavior, with examples drawn from a wide range of situations.

While there are techniques for comparing the goodness of fits of several models to the same dataset (see, for example, Box 4.3), there is no rigid rule that can tell you when to accept or reject the power-law model when fitting data. There is, understandably, fascination with fractal structures, and this has led to uncritical acceptance of power-law models. Moreover, for reasons discussed in Section 4.2, power-law behavior is unlikely in ecology.

---

provides an explanation of how this is done and offers a strong caution that should be heeded when doing so.

The second mathematical preliminary concerns the graphical representation of a probability density function, $g(x)$, or a discrete probability distribution $f(n)$. Such a representation can simply take the form of a plot in which x or n is plotted on the horizontal axis and g or f on the vertical axis. Consider the species–abundance distribution, $\Phi(n)$. A dataset of the abundances of species can be portrayed by plotting integer values of n on the x-axis and the number of species with n individuals on the y-axis as a histogram. But, as mentioned earlier, the sparseness of most abundance datasets implies that such a graph will be a very sparse histogram, with many bars just one species high and with big gaps between bars. For that reason, abundance data are often binned. For example, if one has a list of abundances of species, then the fraction of species with abundances that occur in the interval (n_i, $n_{i+1} = n_i + a$), where i is an integer index and $a > 1$, can be plotted on the vertical axis and with index i on the horizontal axis. That procedure is called "linear binning", but often logarithmic abundance intervals are preferred because the abundance data are spread out logarithmically, and in that case the intervals of abundance would be (n_i, $n_{i+1} = a \cdot n_i$), with $a > 1$. Unfortunately, the shape of the distribution you obtain this way can depend on the way you choose the binning intervals (Bulmer, 1974; Williamson and Gaston, 2005; Gray et al., 2006).

To avoid the ambiguity caused by having to select a binning interval, datasets providing empirical frequency distributions are often graphed in the form of rank–variable relationships. Box 3.4 explains the procedure.

Box 3.4 Rank–variable Graphs

Consider the list of abundances of the plant species in a serpentine meadow plot from a 1998 census (Green et al., 2003). They are listed in tabular form from most to least abundant. There are $S_0 = 24$ species and $N_0 = 37{,}182$ individuals in the dataset. To the left of the column of abundances is a column of ranks. The most abundant species has a rank of 1, and the highest rank, 24 in this case, is the number of species. Note that in cases with two species having the same abundance, they each get ranked.

Table 3.4 Rank-ordered abundances of plants censused in a 64 m² serpentine grassland in 1998. Data from Green et al. (2003); see also Appendix A.

Rank	abundance
1	10792
2	6990
3	5989
4	4827
5	3095
6	1759
7	1418
8	885
9	617
10	272
11	139
12	120
13	112
14	50
15	49
16	30
17	13
18	7
19	6
20	6
21	2
22	2
23	1
24	1

Figure 3.4a plots abundance against rank, and it is not particularly enlightening. But when log(abundance) is plotted against rank in Figure 3.4b, a simple pattern emerges: a straight line.

Let us now see what this implies for the actual probability distribution. Figure 3.4b informs us that abundance, n, is well described by $\log(n) = 9.998 - 0.4264 \cdot r$, where $r =$ rank. Taking the exponential of both sides of this equation, we obtain:

$$n = e^{9.998} \cdot e^{-0.4264\, r}. \tag{3.23}$$

Box 3.4 (*Cont.*)

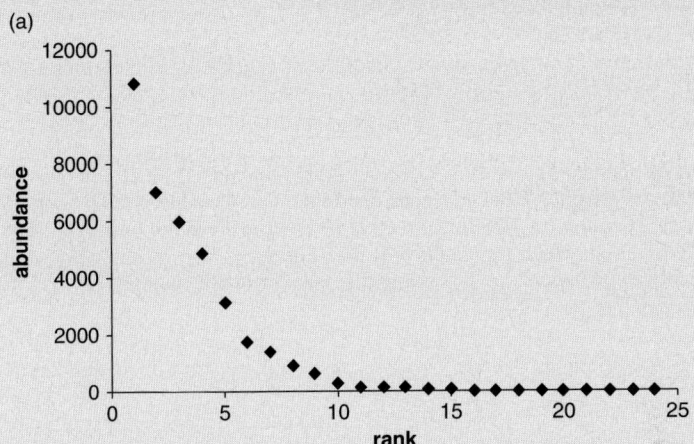

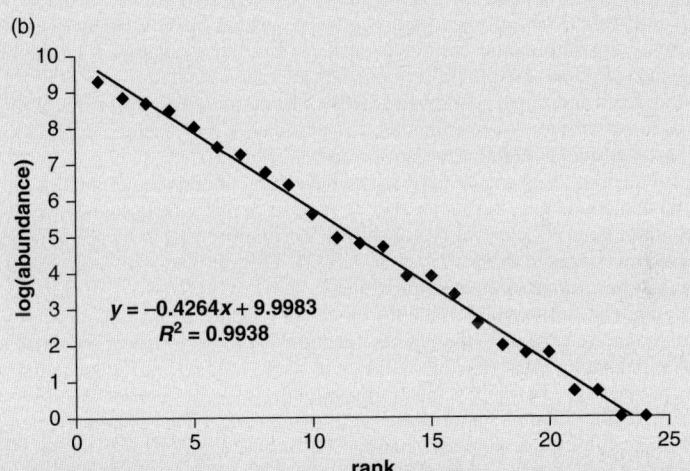

Figure 3.4 Rank abundance graphs for the 1998 serpentine data: (a) abundance versus rank; (b) log(abundance) versus rank.

We can relate the probability distribution, $\Phi(n)$, to this expression by using the formula:

$$\Phi(n) = \frac{-1/S_0}{dn/dr}. \tag{3.24}$$

Box 3.4 *(Cont.)*

Substituting 3.23 into 3.24 and integrating, we get:

$$\Phi = \frac{-1/24}{-0.4264 \cdot e^{9.998} \cdot e^{-.4264 \cdot r}} = \frac{0.0977}{n}. \qquad (3.25)$$

Hence the abundance is well-described by what is called a geometric series distribution. We shall see in Chapter 7 that the actual distribution predicted by METE is of a somewhat different functional form, but for the serpentine plot the predicted form is closely approximated by Eq. 3.25, which we obtained by data fitting.

Eq. 3.24 can be understood by considering the equivalent equation:

$$\int_{n(r+1)}^{n(r)} \Phi(n) dn = S_0^{-1} \int_r^{r+1} dr. \qquad (3.26)$$

Here, r is an arbitrary integer rank, and the negative sign in Eq. 3.24 has been converted to an interchange of the upper and lower limits of integration on the left-hand side of Eq. 3.20. The integral on the right-hand side is just $1/S_0$, which can be thought of as the interval of probability separating adjacent ranked species. But that is exactly what the integral on the left-hand side describes: the difference between the cumulative probability up to $n(r)$ and $n(r+1)$. Each additional species adds $1/S_0$ to the cumulative probability. Equation 3.24 can be used for discrete or continuous distributions if the discrete distribution can be written as a continuous function so that it can be differentiated.

Note also that Eq. 3.24 can be used, as we have here, to derive a probability distribution from rank abundance data but it can also be used to derive a rank–abundance distribution from a theoretical prediction for a probability distribution; you just need to solve for $n(r)$ the differential equation: $dn/dr = -1/(S_0 \cdot \Phi(n))$. That derived rank–abundance relationship can then be compared to an empirical one.

When graphing rank–variable data, the choice of whether to use rank or log(rank) and n or log(n) should be driven by the type of functional form you expect for $\Phi(n)$; often one does know in advance what function will best describe n(*rank*), and so some trial and error in finding a good set of variables is not uncommon.

Figure 3.5a–d give you a better sense of how rank–abundance graphs, plotted on either linear or logarithmic axes, help you identify the form of a probability distribution. Figure 3.5a portrays a straight line on an abundance versus log(rank) graph. Using Eq. 3.24, you can show that this implies $\Phi(n) \sim \exp(-\lambda \cdot n)$, which is an exponential distribution (also referred to as a Boltzmann distribution), where λ is a constant. Figure 3.5b plots the data on linear axes, abundance versus rank, and shows a straight line. Again, using Eq. 3.24, you can see what probability distribution this corresponds to... it is a uniform distribution, $\Phi(n) = $ constant. Figure 3.5c shows three different rank–abundance curves, all plotted as log(abundance) versus rank. The geometric distribution, $\Phi(n) \sim 1/n$, corresponds on those axes to an exactly linear graph. On those axes, the signature of the logseries distribution, $\Phi(n) \sim (1/n) \cdot \exp(-\lambda \cdot n)$, is that it resembles the geometric distribution but there is an upturn at high abundance (low rank).

Box 3.4 *(Cont.)*

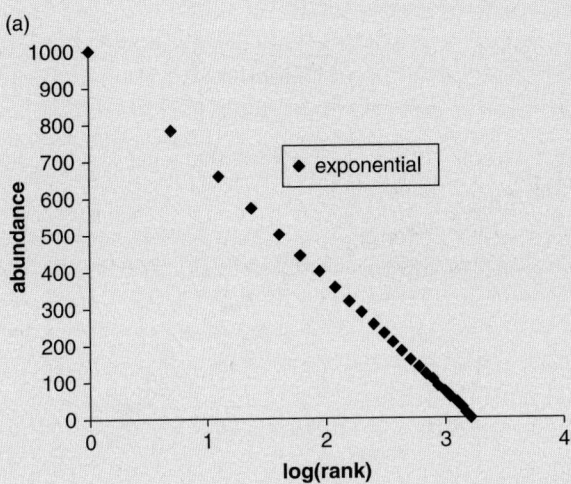

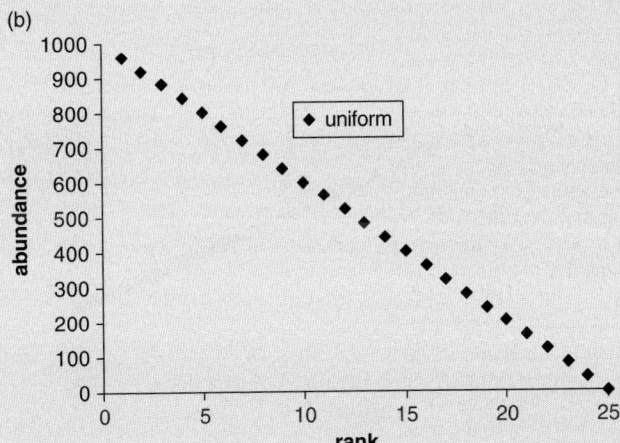

Figure 3.5 Four ways to plot ranked abundance data, illustrating how the functional form of the species abundance distribution can be revealed by examining which type of plot looks simplest. (a) An exponential distribution, $p(n) \sim e^{-cn}$. (b) A uniform distribution, $p(n) \sim$ constant. (c) A geometric distribution, $p(n) \sim 1/n$, a logseries distribution, $p(n) \sim e^{-cn}/n$, and the lognormal distribution. (d) A Cauchy or Zipf distribution, $p(n) \sim 1/n^2$.

Box 3.4 (*Cont.*)

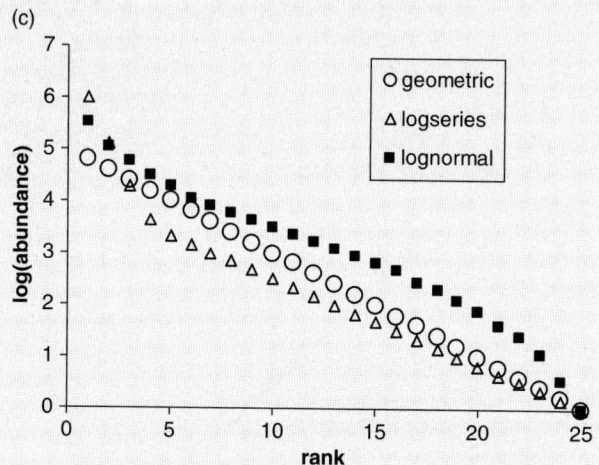

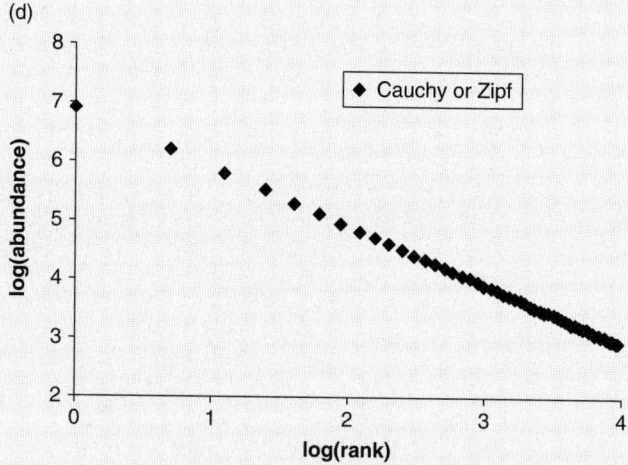

You can also see in the figure the characteristic signature of unimodal (meaning they first rise and then fall with abundance) distributions, such as the lognormal, $\Phi(n) \sim \exp[-(\log(n) - a)^2/b]$. The way to see that a rank versus either abundance or log(abundance) graph corresponds to a hump-shaped distribution is to note that where the rank–abundance graph is most flat, $\Phi(n)$ must be relatively large. The reason is that the range of abundance values for which the rank–abundance graph is most flat must be

Box 3.4 *(Cont.)*

relatively small; many ranks means many species, and thus many species possess a narrow range of abundances, which is exactly what it means for $\Phi(n)$ in that range to be large. The reverse symmetry of the low- and high-rank ends of the lognormal graph in Figure 3.5c, with log(abundance) on the vertical axis, is a consequence of the lognormal distribution being a symmetric distribution (in log(n)) about its maximum.

Finally, Figure 3.5d plots a hypothetical discrete distribution on log(abundance) versus log(rank) axes. Here a straight line indicates a power-law $\Phi(n) \sim n^{-a}$; the slope of the log (abundance) versus log(rank) graph can be shown to be $(1-a)^{-1}$ in that case (Exercise 3.2). The particular case shown in Figure 3.5.d is for $a = 2$, and it yields a rank abundance curve of the form abundance $\sim 1/$rank, which is also called a Cauchy or Zipf distribution (Zipf, 1949). This distribution has been of particular interest in economics and demography, for reasons suggested in a useful article by Gabaix (1999).

While I have illustrated the above with abundance as the variable of interest, the same ideas apply to distributions of metabolic energy or body mass or number of modal links in a web, or any other variable of interest.

3.4.1 Species-level spatial-abundance distributions: $\Pi(n|A, n_0, A_0)$

Figures 3.6a–c present species-level spatial abundance data from the serpentine plot for three species of plants at the spatial scale $A = \frac{1}{4}$ m$^2 = A_0/256$. I have plotted the numbers of $\frac{1}{4}$ m^2 cells that contain n individuals versus n, or equivalently $256 \cdot \Pi(n|1, n_0, 256)$. The species were deliberately chosen to show a range of abundances.

The graphs illustrate some ubiquitous patterns, seen in plant-distribution data from many habitats, although exceptions certainly exist. First, note that the graphs show a general pattern of monotonic, or nearly monotonic, decline of Π with increasing n. The central tendency of these graphs suggests that there are more cells with 0 individuals than with 1 individual, more with 1 than with 2, and so on.

The data in Figure 3.6 contrast the observed cell occupancy distribution with the binomial distribution that results from a model (Coleman, 1981) called the random placement model (RPM). The RPM assumes that all individuals of a species are randomly located on the landscape. To get a sense of what random placement means, imagine dropping 100 seeds from a great height on to a plot; wherever a seed lands, an individual of an $n_0 = 100$ species will be located. The actual shape of $\Pi(n|A, n_0, A_0)$ in the RPM, the binomial distribution, is derived in Chapter 4. If $n_0 > A_0/A$, the binomial distribution for $\Pi(n|A, n_0, A_0)$ is hump-shaped rather than monotonically falling with n. Note that the condition $n_0 < A_0/A$ corresponds to there being more individuals than there are cells to place them in, so at least some cells must have more than one individual. If $n_0 < A_0/A$, then the binomial distribution prediction for $\Pi(n|A, n_0, A_0)$ is monotonically decreasing with n.

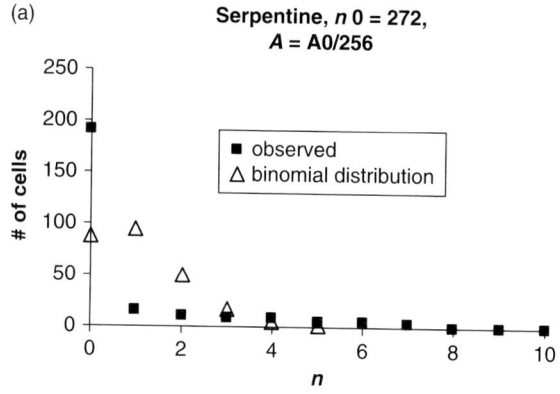

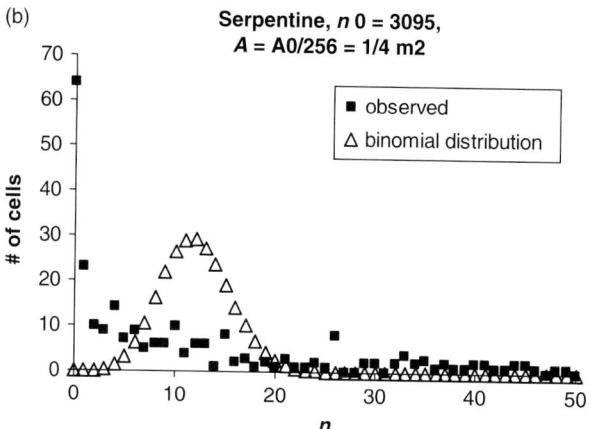

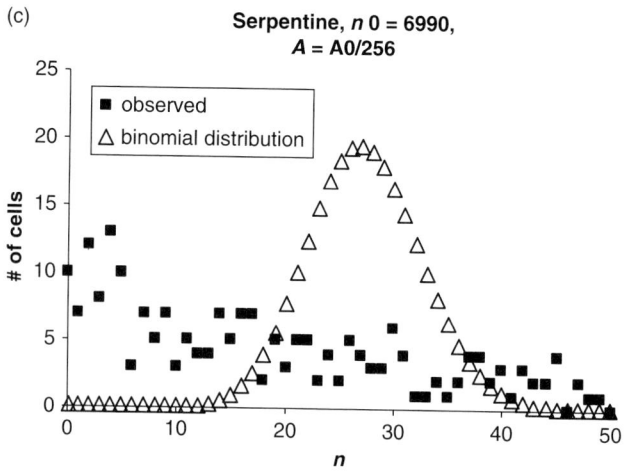

Figure 3.6 Fraction of cells of area ¼ m² within a 64 m² serpentine grassland plot (Green et al., 2003) that contain n individuals for a plant species with a total of: (a) 272 individuals; (b) 3095 individuals; (c) 6990 individuals in the whole plot. The data are compared to the binomial distribution predicted by the Coleman (1981) random placement model.

For $n_0 A/A_0 \ll 1$, it is difficult to distinguish empirical spatial distribution patterns from the binomial distribution, but at larger values, where the binomial distribution is hump-shaped, empirical spatial distributions usually indicate a monotonically falling pattern, as seen in the figures. However, for a fixed value of A_0/A, as n_0 increases there is a tendency for the falling distribution to flatten.

3.4.2 Range–area relationship: $B(A|n_0, A_0)$

Recalling that we are defining this metric for range size using a box-counting method that has its origins in fractal analysis, it is reasonable to graph $B(A|n_0, A_0)$ in such a way as to allow us to determine if species' distributions are in fact fractal. As discussed in Appendix B, a species would occupy its range in a fractal pattern if a graph of log(B) vs. log (A) is a straight line, or in other words, $B \sim A^y$, where y is independent of A. Exercise 3.14 asks you to do this for plant species from the serpentine plot that were selected in Figure 3.6.

Log(range-size) versus log(area) graphs for most species in most habitats are not straight, but rather exhibit negative curvature (Kunin, 1998; Green et al., 2003) that is considerably greater than the curvature in the data shown in Figure 3.3.

3.4.2.1 A note on the nomenclature of curvature

Describing curvature can be confusing. Some authors refer to convex and concave curvature but I find that confusing. When I write "negative curvature" I mean that the slope is decreasing as area increases. Equivalently, the second derivative is negative. These particular curves are rising, so a decreasing slope means they rise more slowly as area increases. But if a function were descending as the abscissa variable increased, so that the slope was negative, negative curvature would still mean a decreasing or even more negative slope; in other words, the curve would be descending ever steeper as the abscissa variable increased.

3.4.3 Species-level commonality: $C(A, D|n_0, A_0)$

The species-level commonality metric is observed to be an increasing function of A and is often a decreasing function of distance, D. Its increase with A is obvious; the probability of finding an individual of a particular species in a big cell is greater than it is in a small cell. So the probability of finding it in both of two cells increases with cell area. The dependence on distance is less obvious. Several arguments suggest that C might decrease with distance. For example, if a species that occurs within a large biome is preferentially found in patches of suitable habitat, such as steep north-facing slopes, then it would be more likely that two patches on such slopes will both contain individuals of that species, whereas a patch on that slope and another patch further away on a south-facing slope will not. As a consequence, C would decrease with distance. Moreover, if plants rarely disperse seeds large distances from the parent, then, again, C will decrease with distance.

But on the other hand, there are ecological forces at work that act to reduce the likelihood that an offspring of an animal or a surviving germinant will be located near its parent. For plants, this is called the Janzen–Connell effect (Janzen, 1970; Connell, 1971) mentioned in Section 3.3.3. The argument is that if the individuals in a plant species are clustered somewhere, then herbivores that are adapted to eating those individuals will proliferate in that cluster and the cost in fitness to the plant will be great. Hence, the argument goes, evolution has favored dispersal mechanisms that at least disperse seeds far enough from the parent to avoid creating an attraction to herbivores. Another reason for a possibly increase of C with distance is that fledgling birds and migrating young mammals forced to leave their birth burrow may not be able to raise their young too close to their birthplace because of food-supply limitations. Putting this all together, we might predict that C should increase for small values of D and then decrease at larger distances.

Relatively few studies have systematically examined the D-dependence of C. One extensive survey (Nekola and White, 1999) found that a pervasive, but not universal, pattern is for C to decrease with distance, which is why the D-dependence of C is sometimes referred to as distance-decay. Their study found that C was reasonably well-described by an exponentially decreasing function of distance. There is evidence in the tropics for the Janzen–Connell effect, but see Hyatt et al. (2003). More systematic surveys would be useful.

To understand how distance-decay depends upon overall population density, n_0/A_0, Condit et al. (2000) examined the "O-ring" measure of aggregation defined in Section 3.3.3. They evaluate it with tree-location data from the Smithsonian Tropical Forest Science (STFS) 50-ha plots. Using linear rather than logarithmic axes appears to best display patterns, with distance, D, on the horizontal axis and the metric Ω on the vertical axis. By plotting species of different abundance on the same graph, a visual sense of how clustering depends on total abundance in the 50-ha plot is obtained. A pervasive pattern was revealed for the larger trees (dbh $\geq$ 10 cm); for nearly all species at all six STFS sites, clustering declined with distance, clustering was greater than expected under random placement, clustering was greater for rare than for common species, at least at relatively small D (less than about 100 m). In other words, Ω declined with distance, it was generally > 1, and for $D < 100$ m, Ω was a declining function of the abundance of the species. At small distances (< 10 m) Ω values for many species indicated aggregation was 5 to 30 times as high as expected under randomness.

Condit et al. conclude that there was some evidence at some sites for topographic or other habitat related effects on aggregation, but that dispersal limitation explains more of the observed patterns. Did species' traits (other than abundance and dispersability) matter? The evidence is somewhat equivocal. At some, but not all, sites, aggregation intensity differed between understory and canopy species, and at some sites there was a non-significant difference between animal- and non-animal-dispersed species. In two of the plots, dipterocarps were more aggregated than non-dipterocarps. Site differences in the D-dependence and magnitude of Ω appear to be as large as within-site differences for species of nearly the same abundance, with both effects smaller than the effect of abundance itself.

Analysis of clustering patterns as systematic as the Smithsonian tree analysis has not been carried out for animal distributions, nor for many other plant datasets.

3.4.4 Intra-specific distribution of metabolic rates: $\Theta(\varepsilon|n_0)$

We will return to this metric when we discuss mass–density relationships under Section 3.4.11.

3.4.5 Intra-specific distribution of dispersal distances: $\Delta(D)$

Some general considerations about dispersal distributions were discussed briefly in Section 3.4.3, but there are few datasets that allow us to actually examine the shape of $\Delta(D)$. There are several reasons for this. First, for plants, measurements of seed fall have been carried out at various distances from trees, but it is difficult to trace a seed back to its parent, and thus obtain reliable information about the value of D for a given seed. The use of genetic markers to identify the parent seed-producer has begun (Hardesty et al., 2006) and this technique, though currently expensive, offers much promise. Other efforts have focused on seed fall distributions from individual parent trees that happen to be located far from conspecific individuals or on seed fall from trees in open spaces at the edge of a forest. But because of wind conditions and possible differences in the behavior of seed-transporting animals in these circumstances, the dispersal distributions obtained may be specific to these specific situations.

A second impediment to measuring $\Delta(D)$ for plants, stems from the fact that it describes not the actual fallout pattern of seeds from a parent, but rather the spatial pattern of reproductively viable offspring from a parent. This cannot be determined simply with seed rain traps; in a forest, one would either actually have to connect together, in a network, the genetic relatedness of the conspecifics of different ages, or find parent trees that are sufficiently isolated that one can reasonably assume that nearby offspring sprang from that parent.

Despite the difficulties obtaining reliable measurements of the shape of $\Delta(D)$, some qualitative insights have been obtained. For plants, there is a general consensus that at small scales the distribution is roughly flat, or perhaps slightly rising, due to the Janzen–Connell effect (Section 3.4.3). Then, at intermediate distances, it falls off rather rapidly, perhaps like a Gaussian distribution ($\sim\exp(-D^2)$). However, at very large distances, there is some evidence that occasional dispersal events occur at a rate that greatly exceeds the prediction based on extrapolating the Gaussian. Such events could occur if animals occasionally deposit seeds at great distances or if very infrequent wind storms do the same.

One simplified fitting function that has been used to describe the overall pattern is a Gaussian distribution adjoined at large distances to a so-called fat-tailed distribution. An example of the latter is the Cauchy distribution $\Delta(D) \sim 1/D^2$. Fat-tailed distributions are functions with the property that their moments do not all exist because at large values of their independent variable (D in this case), the distributions fall to zero too slowly. The Gaussian distribution is not fat-tailed because it falls to zero exponentially, thus guaranteeing the existence of the expectation values of all powers of D and thus all of the moments of the distribution.

I know of no systematic empirical evaluation of the dependence of dispersal distance on abundance. One theoretical study (Harte 2006) derived the dependence of the dispersal distribution $\Delta(D|n_0)$ on species' abundance that was necessary for consistency with a particular model of the species-level spatial abundance distribution $\Pi(n|A, n_0, A_0)$. Because of the importance in ecology of dispersal distributions, and the difficulty in directly measuring them, I return in Section 7.9 to the topic of inferring Δ from Π.

3.4.6 The species–abundance distribution: $\Phi(n|S_0, N_0, A_0)$

Paralleling what we said about the function $\Pi(n)$ in Section 3.4.1, two qualitatively distinct shapes of SADs have been discussed. One is a monotonically decreasing function of n, and the other is hump-shaped or "unimodal." Remarkably, even for a specified dataset, arguments arise over which of these forms best describes the data. How can such ambiguity exist?

Recalling Box 3.4 and the discussion preceding it, graphs of the SAD generally take one of two forms. Sometimes the probability distribution $\Phi(n)$ is directly plotted, with the fraction of species with n individuals expressed on the y-axis and n or, more often, log(n) on the x-axis. The use of log(n) on the x-axis allows a more uniform distribution of data points along that axis. The reason stems from the fact that often there are many species with quite low abundances and a very few with very high abundance; hence, plotting $\Phi(n)$ against n rather than log(n) leads to a bunching up of many data points at small n, so that it becomes difficult to discern their pattern. When the number of species observed with abundance n is plotted against n or log(n), and the distribution appears unimodal (that is, it has a peak value), the distribution is generally not symmetric. In particular there tend to be more rare species than would be the case if the distribution were symmetric about the mode. In contrast, the lognormal distribution would be symmetric when the x-axis is log(n). As we discussed in Box 3.4, however, it is generally most instructive to avoid the ambiguity caused by having to select a binning interval, and instead plot the SAD in the format of a rank-abundance graph.

The three functions shown in Box 3.4, Figure 3.5c, cover most of the patterns observed in, or discussed in relation to, abundance data. An important exception is the zero-sum multinomial distribution that results from the Hubbell neutral theory

(Hubbell, 2001; McGill, 2003; Volkov et al., 2003). The logseries distribution, or a distribution like the logseries but with some downturn at large rank, arguably describe the majority of datasets. On a log(abundance) versus rank graph, the asymmetry mentioned above in unimodal distributions is such that the upturn at low rank is generally not a reverse image of the downturn at high rank (note the symmetry of the lognormal distribution in Figure 3.5c). Arguments about which function best fits abundance distribution data are in some cases not worth fighting because of concern about whether the high and low rank ends of the distribution have been adequately measured. It is easy to miss very rare species when conducting a census, and anyone who has tried to count the number of blades of grass in a patch of meadow or of birds in a sky-filling flock of blackbirds knows how hard it is to estimate abundances of common species.

The issue of incomplete censusing hampers our ability to generalize too much about prevalent patterns in the species–abundance distribution. An interesting analysis by Loehle (2006) revealed one facet of the problem. He created a large dataset numerically from a Fisher logseries distribution (Figure 3.5c) and then sampled from it randomly. The samples looked more like the lognormal distribution also shown in that figure. Thus subsampling from a monotonically falling $\Phi(n)$ can generate a unimodal $\Phi(n)$. Some datasets are relatively immune from this problem however. For example, the study resulting in the serpentine abundance distribution, shown in Figure 3.4, was carried out thoroughly and on a small enough plot, so that one has confidence that nearly every individual plant was included in the census. In contrast, the $\Phi(n)$ for the Smithsonian 50-ha plots is taken from a census strategy in which only trees with greater than 1-cm diameter-at-breast-height are included. The implications for the SAD of having such a cutoff are not well understood. And then other datasets, such as breeding bird censuses, are clearly only approximations to what is really out there, in somewhat the same sense that the lognormal distributions Loehle obtained from subsamples are approximations to the numerically generated logseries distribution.

One compelling insight into the conditions under which unimodal versus monotonically-declining SADs are observed was first enunciated by Kempton and Taylor (1974). They carried out extensive censuses of moths in different local habitats at Rothampsted, a site in England where for over 100 years important research in ecology has been carried out. Local habitats vary across fields in which different agricultural practices, including abandonment of cropping and return to wild vegetation, have been deployed over the years. In most of the large plots they censused, abundance distributions were consistent with the logseries function, but in plots where crop abandonment had most recently occurred, and the plant communities were in the greatest state of flux, unimodal distributions were observed. The lognormal function provided a reasonable distribution of the SAD in those disturbed plots.

3.4.7 The species–area relationship: $\bar{S}\,(A|N_0, S_0, A_0)$

Many functions have been suggested to describe SARs of all types: island or mainland, nested or disjunct areas. Gleason (1922) first proposed the logarithmic function: $\bar{S} \sim \log(A)$, Arhennius (1921) the power-law function, $\bar{S} \sim A^z$, and Lomolino (2001) a sigmoidal function $\bar{S} \sim A^z/(b + A^z)$. Another hypothesized qualitative form for the SAR is a "triphasic" shape, in which, as area increases, slope on a log–log graph first decreases and then increases at larger scale. The power-law model, however, has become the default assumption in most SAR investigations. Because testing the goodness of a power-law fit is easiest on log–log graphs (Box 3.3), species–area data are usually plotted as $\log(\bar{S})$ versus $\log(A)$.

Using log(area) for the x-axis variable is also useful because the adjacent areas measured in nested datasets obtained from gridded plots generally differ by a factor of two; hence on a log scale the data are equally spaced along the x-axis. For example, in the serpentine plot, the smallest cell area that can be used in the SAR is ¼ m². Adjoining two of these, gives cells of area ½ m². Adjoining two of those cells gives 1 m² cells, and so forth. Hence the areas that would normally be analyzed are ¼, ½, 1, 2, 4, 8, 16, 32, 64, 128, and 256 m². Because these areas are spread out logarithmically, graphing the data on a log scale gives uniform spacing of data on the x-axis, and that makes it easier to visualize. Of course one could take, say, 24 adjacent ¼ m² cells and make 6 m² cells, but those cells would not fill, without overlap, the entire 8 m × 8 m plot.

Many researchers have examined the shapes of SARs, searching for systematic patterns. The most frequently asked questions are:

Do island SARs systematically differ, in fitted slope and goodness of fit, from mainland SARs, and do independent area SARs systematically differ in those same respects from nested SARs?
How successful is the power-law model; does any simpler function provide a better model, and in what direction do empirical SARs differ from the power-law model?
How does fitted slope and goodness of fit vary with the spatial scale of the study?
Is there a latitudinal gradient in the fitted slope and goodness of fit?
Do fitted slopes and goodness of fit vary systematically across taxonomic groups; does typical body size of the organisms under study correlate with fitted slope?
How does fitted slope and goodness of fit vary with habitat type?

In a wide-ranging and well-documented review of the SAR literature, Drakare et al. (2006) have provided what are probably the most reliable overall answers to these questions. Their analysis, which examined 794 SARs, resulted in the following generalizations:

Nested SARs tend to have steeper slopes than independent-area SARs on islands and mainlands, and the power-law fit for nested SARs tends to be better than for independent-area SARs. Island SARs did not differ in fitted slope from mainland SARs.

Most SARs exhibit negative curvature on log–log axes; the power-law model and the Gleason (logarithmic) model are roughly comparable in their success.
Fitted power-law slopes tend to decrease with increasing scale range of analysis. Scale range does not affect the goodness of the fit but the diminishing number of data points at large scales makes it difficult to interpret goodness measures. Evidence for a triphasic SAR, in which slope first decreases from small to intermediate scale and then increases at larger scale (see below), was scant.
Fitted slopes tend to decrease with increasing latitude. In addition, of course, the SAR comparisons confirm the widely noted decrease with latitude of species' richness at any scale.
There is a strong, positive correlation between typical body sizes of organisms in the taxonomic group being studied and fitted power-law slopes. Otherwise there is no systematic dependence of slope on taxonomic group.
Only minor differences are found between fitted slopes for aquatic and terrestrial SARs, with the latter having slightly larger slopes. The Gleason function fits terrestrial SARs better than it did aquatic SARs. Forested habitats have higher fitted power-law slopes than non-forested habitats, and marine habitats have higher slopes than freshwater aquatic habitats.

Some of these conclusions are contrary to earlier reports in the literature. For example, Rosenzweig (1995) concluded that fitted slopes to island SARs tend to be steeper than those for mainland SARs. Lyons and Willig (2002) reported a trend of fitted slope with latitude that was opposite to the Drakare et al. finding, while Rodriguez and Arita (2004) found a mixed pattern for latitudinal dependence, with bats exhibiting an increase of slope with latitude and non-volant mammals a decrease.

One robust generalization that emerges from numerous reviews of SARs is that no one simple function such as the ones mentioned at the beginning of this subsection appear adequate to describe the actual shape of the SAR, in all of its manifestations across census designs, spatial scales, habitats, latitude, and body sizes, and taxonomic groups. In Chapter 8 I will show that the SAR predicted by METE, while not expressible as a simple function, does consistently match the patterns highlighted by Drakare et al.

In the third bulleted item above, the triphasic SAR was mentioned. This postulated shape for SARs is illustrated in Figure 3.7. The portion of the curve at small to intermediate area is typical of SARs observed widely at that scale range; on log–log plots, the pervasive pattern is that slopes decrease with area. As area increases, however, and distinctly different habitats are captured in the census, the triphasic species' richness will increase with area because different habitats have different species. Qualitatively this explanation is reasonable. However, actual nested SAR data over large enough areas to see this effect are not available, and so published evidence for triphasic curves comes from studies in which there is inconsistent data sampling across the wide range of spatial scales needed to see the triphasic behavior.

Thus the phenomenon is entirely reasonable, but available data are not sufficient to give us quantitative insight into its detailed form.

One other trend is discussed in the literature, but not by Drakare et al. The ratio of the number of species found in two patches of the same area within a biome can depend upon their relative shapes. Generally, long, skinny patches contain more species than do square ones of the same area (Kunin, 1997; Harte et al., 1999b). This is because long, skinny patches are more likely to include species' turnover across the increasingly varied habitats often encountered over larger and larger distance intervals.

Figures 3.8a–c shows three SARs: data from the serpentine plot, from a 50-ha, wet tropical forest plot at Barro Colorado Island in Panama, and from a 10-ha dry tropical forest plot at San Emilio, Costa Rica. Note that a straight-line fit to the log–log plot of the Serpentine data looks impressive, whereas the other two SARs exhibit considerable curvature. Comparing the R^2 for the straight-line fit to the serpentine SAR with that for the data in Figure 3.3 in Box 3.3, you might be tempted to say that the difference is unimportant: 0.9993 vs. 0.9899, but in fact the difference in these two straight-line fits is large. One reason is that it is $1 - R^2$, the unexplained variance, that provides a measure of the badness of the fit. By this measure the two graphs differ in the ratio $(1 - 0.9993)/(1 - 0.9899) \sim 0.07$ meaning that the data in Figure 3.3 are in some sense 14 times less well fit to a power law than the data in Figure 3.8a. There is a second important difference in the two graphs. There is clearly systematic curvature in Figure 3.3. Had the data points deviated randomly from the straight line, a power-law model fit with an $R^2 = 0.9899$ would have more credibility than does the fit in the figure.

In Chapter 4 I discuss a different method than comparing R^2 values for evaluating the comparative goodness of different functions arising from different models. Here we simply point out that some nested SARs appear rather linear on log–log plots, while others do not. An explanation of the ecological circumstances in which we should or should not expect the power-law model to fit nested SARs well, will be advanced in Chapter 7.

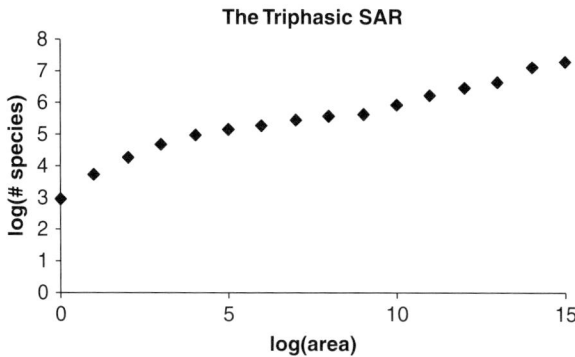

Figure 3.7 An example of a (hypothetical) triphasic species–area relationship.

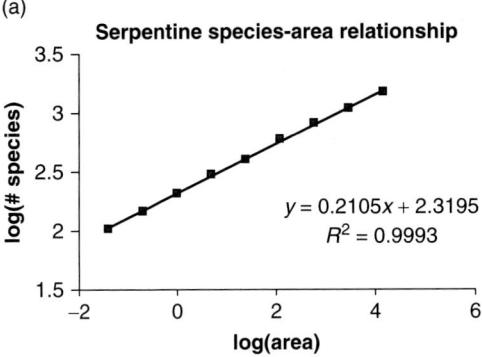
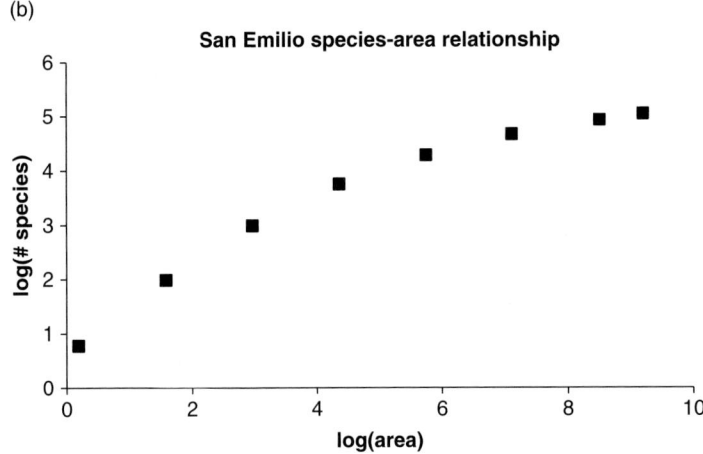
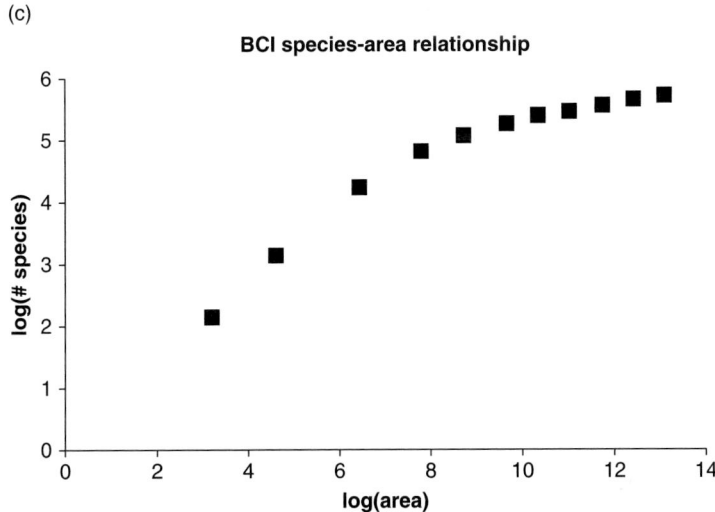

Figure 3.8 Observed species–area relationships for: (a) the serpentine grassland plot (1998 data); (b) San Emilio 10-ha dry tropical forest plot (Enquist et al., 1999); (c) the BCI 50-ha plot (Condit, 1998; Hubbell et al., 1999, 2005).

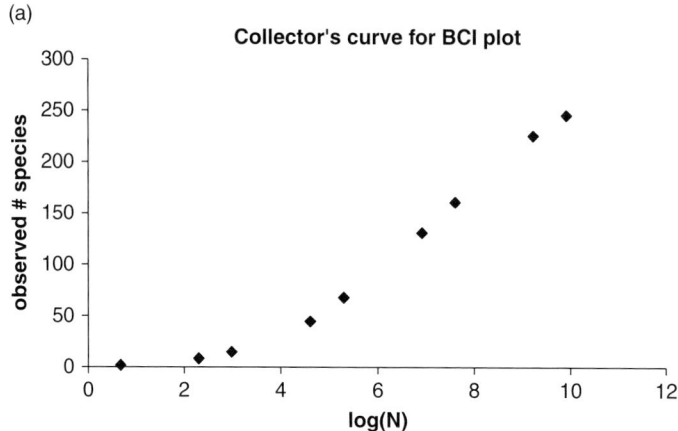

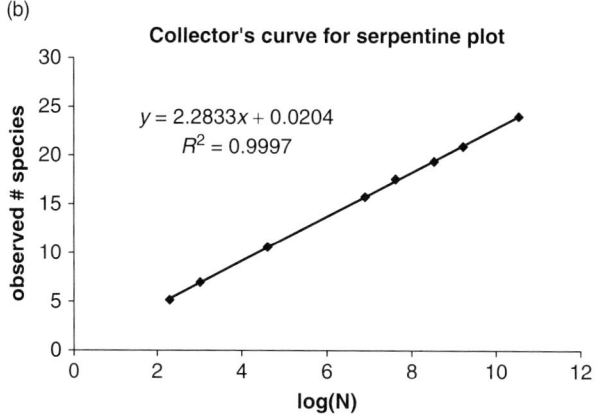

Figure 3.9 Collector's curves for plant data from: (a) the Barro Colorado Island 50-ha plot (cite); (b) the 64-m² serpentine grassland plot (Green et al., 2003).

Related to species–area curves are collector's curves (sometimes called species-accumulation curves and discussed in Section 3.3.7). As I showed earlier, they are derived from knowledge of the SAD alone. Typically they are plotted with log (number of individuals sampled) on the x-axis and cumulative number of species found on the y-axis. Often these plots appear roughly linear, at least at large sample size, implying that the number of species collected grows as log(number of individuals sampled). Two examples of collector's curves are shown in Figures 3.9a, b. For each plot, the collector's curve was obtained from the empirical species–abundance distribution by averaging over 1000 Monte Carlo runs; in each run, the individuals are drawn, one at a time, from the individuals' pool until the pool is drained. The sampling was carried out without replacement, meaning that the sampled individuals are not returned to the individuals' pool during each run.

Note that the serpentine collector's curve is a straight line on those axes, so that $S \sim a + b \cdot \log(N)$, as is the rank versus log(abundance) graph for the serpentine SAD shown in Figure 3.4b ($\log(N) \sim a - b \cdot \text{rank}$). In Chapter 4 we shall see that these two logarithmic relationships are connected. In contrast, the BCI curve, while asymptoting to a linear relationship between S and $\log(N)$, exhibits at small N a more linear relationship between S and N. This, we shall see, is no accident.

3.4.8 The endemics–area relationship: $\bar{E}(A|N_0, S_0, A_0)$

EARs are usually plotted the same way as are SARs. Figure 3.10a and b show the EAR for the serpentine and the BCI plots, and the patterns exhibited there are fairly typical. At small scales the behavior is fairly consistent with a straight line with slope just a little greater than 1, while at larger scales, the slope increases. Exercise 3.6 asks you to explain why complete nested EARs, plotted on log–log axes, must always have a slope that is ≥ 1, in contrast to SARs that must have slopes ≤ 1.

3.4.9 Community-level commonality: $X(A,D|N_0, S_0, A_0)$

This metric of community-level commonality has been studied primarily for plant data, although some bird censuses do provide the information needed to evaluate it. X is often plotted the same way that C is, using $\log(X)$ versus either $\log(D)$ or $\log(A)$ graphs. Nekola and White (1999) examined the distance dependence of X in North American spruce-fir forests over distances up to 6000 km. They found that an

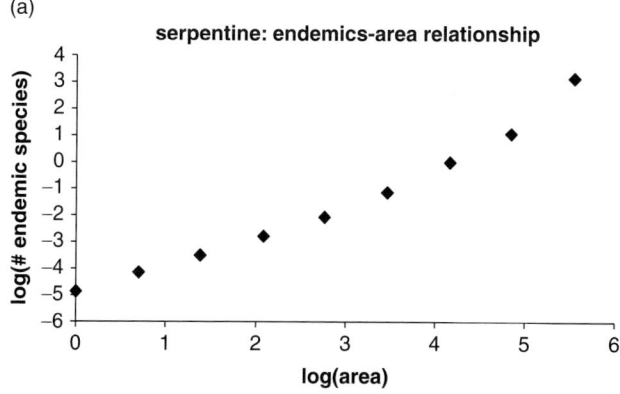

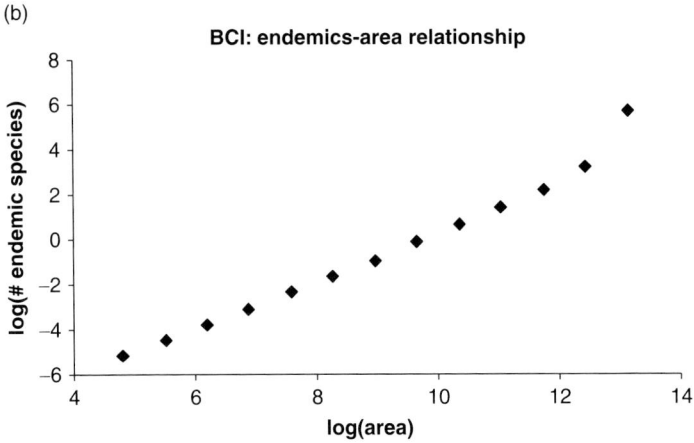

Figure 3.10 Observed endemics–area relationships for: (a) the serpentine grassland; (b) BCI.

exponential decay function, $X \sim \exp(-cD)$, captured the central tendency of all their North American forest data.

Condit et al. (2002) examined this metric for trees in the BCI 50-ha plot, where distances up to about 1.4 km are accessible. They also looked at data over much larger spans of tropical forest by examining the fractions of tree species shared between 1-ha plots at sites in Panama and western Amazonia, with inter-plot but within-site distances extending up to 100 km. The Condit et al. (2002) analysis found that the fraction of species shared by two plots, graphed as X versus D, generally falls steeply at small distances and then tends to flatten at larger D. A single exponential function did not adequately describe their data, however, and they also concluded that the data contradicted a neutral theory (Hubbell, 2001) prediction. Distance-decay was steeper at the Panama site than in western Amazonia, and the number of species shared between the two sites (~ 1400 km apart) was very low. Condit et al. conclude that no simple function appears to capture the observed patterns, and they attribute the difference in the shape of X within the Panama and western Amazonia sites to greater habitat heterogeneity in the former.

Another analysis (Krishnamani et al., 2004) for forest trees in the Western Ghats in India examined X over distance ranging up to ~ 1000 km, with $A = \frac{1}{4}$ ha. They found that a single power-law model failed to capture the distance dependence of X over distances from a few tens of meters up to 1000 km, with the slope declining with inter-plot distance.

The slowest rates of "distance-decay" have been observed for microbial data (Green et al., 2004; Horner-Devine et al., 2004). They examined this metric for microbial samples from soils, with D-values ranging from less than a meter to ~ 100 km, and an A-value less than the area of small soil cores but indeterminate because of the way that microbial genetic material is extracted from soil core samples.

Power-law fits to graphs of log(X) versus log(D) yielded slopes of order –0.1, which imply a much lower rate of distance-decay, plotted as log(X) versus log(D), than is typically found for plants (Harte et al. 1999b; Krishnamani et al. 2004; Condit et al. 2002).

Rarely has X been examined empirically as a function of both independent variables A and D; one such study, however, on subalpine meadow flora, examined plots ranging from 0.25 m^2 to 16 m^2 in area and inter-plot distances ranging from plot dimension to ~ 10 km (Harte et al., 1999b). X increased with A at fixed D and decreased with D at fixed A; fitting power-law behavior separately to each of these showed that the rising A-dependence was considerably steeper than the falling D-dependence. To a good approximation, the D-dependence was decoupled from the A-dependence, in the sense that the graphs of log(X) versus log(D) for different values of A were simply shifted up or down on the log (X) axis but their shape did not depend on A.

3.4.10 Energy and mass distributions and energy– and mass–abundance relationships

Because masses are easier than metabolic rates to measure, the mass metrics in Tables 3.2 and 3.3 have been the primary focus of investigation. And of all the metrics, the ones that have received nearly all the attention have been the mass–abundance relationships (MARs) listed in Table 3.3. Patterns in the MARs are revealed by graphing log(abundance) on the vertical axis and log(mass) on the x-axis.

Marquet et al. (1990) conducted one of the first careful analyses of the mass–population density relationship. Their study, in intertidal animal communities, showed evidence for a power-law relationship density ~ mass$^{-0.77}$. White et al. (2007) provide a comprehensive summary of selected available data for several taxonomic groups. They conclude that plots of data for any of the four measures of the MAR are consistent with abundance decreasing with increasing average mass of individuals. The IMD decreases fastest, and fitted to a power law, the data in White et al. show abundance ~ mass$^{-3/2}$. The CCSR decrease second fastest, with roughly an inverse linear relation between total abundance and average mass. White et al.'s plot of GMDR data indicates density decreasing as mass$^{-3/4}$, and the LSDR shows the most scatter and a very shallow relationship between density and mass.

The empirical finding for the GMDR, with its inverse ¾-power relationship between density and mass, is called the Damuth rule (Damuth, 1981) and it is related to a phenomenon called energy-equivalence, if metabolic energy follows the MST prediction: $\epsilon = c \cdot m^{3/4}$. The concept of energy-equivalence is this: suppose that within some region every species has a density that is inversely proportional to the metabolic rate of an individual in that species. Then every species would consume the same amount of energy in the region. In other words, if $\epsilon = c \cdot m^{3/4} \sim 1/n$ for each species, then metabolic rate per individual times the number of individuals, $\epsilon \cdot n$, is the same for every species. Each species utilizes the same share of total community energy, E_0.

The results of White et al. present an interesting challenge. They show that it is the GMDR that is consistent with energy-equivalence, but it is not at all clear why that metric and not one of the other measures of the MAR should reveal the $-\frac{3}{4}$ scaling relation between abundance or density and mass. Possible explanations are reviewed in their paper, but their conclusion is that this remains a puzzle. We shall return to this issue in Chapter 7.

Finally, in a review of metabolic scaling theory, Brown et al. (2004) include a graph (Figure 3.11) showing evidence across a wide variety of taxa for an inverse ¾ power relationship between abundance and body mass.

3.4.11 $\Lambda(l|S_0, L_0)$

Like abundance distributions, link distributions could be plotted either as histograms of binned data or as rank–link graphs. For the same reasons that we discussed for the SAD, the rank–link graphs generally provide more insight. Williams (2010) examined the distribution of links in a large collection of food web datasets and showed that for about half the webs the exponential distribution, $\Lambda(l) \sim \exp(-b \cdot l)$, described $\Lambda(l|S_0, L_0)$ while the rest deviated somewhat, with no prevalent pattern discernable in the deviations.

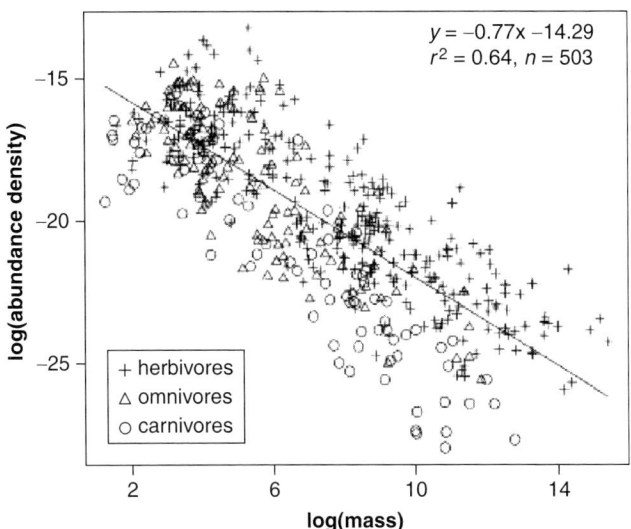

Figure 3.11 The log of population density plotted against the log of average body mass for numerous species of animals. The slope of –0.77 suggests a power-law relationship of the form population density $\sim (\text{mass})^{-0.77}$. From Brown et al. (2004).

3.4.12 ε(m)

As with many other metrics, log–log graphs are generally used to portray the dependence of basal metabolic rate of animals on body size or mass, and the reasons are similar. One is the roughly six orders of magnitude range in body mass, from small shrews to large whales, across just the mammals. And if zooplankton and shrews and whales are compared, the range expands by another seven orders of magnitude. Such huge ranges of mass are best plotted on a logarithmic scale to avoid obscuring the variations among the small creatures. A second reason is that the leading theory relating metabolism to body size produces a power-law relationship, and the validity of that relationship is best evaluated by looking at the linear regression of the log–log graphs of the data. One example of such a graph is shown in Figure 3.12. Putting aside arguments about whether or not the scaling exponent is exactly ¾, the data show clear evidence for a systematic patterns relating body mass and metabolic rate.

There is a vigorous and ongoing debate over the validity and generality of the ¾-power scaling law (Eq. 1.1). Some have accepted the argument that the ¾ slope is a best fit to enormous amounts of data, and while there is scatter around that best fit, the MST (West et al., 1997) is still a useful theory. Others have argued that the best fitting slope is not exactly ¾ and so the theory is fundamentally wrong. Still others have argued that the ¾ fit only describes some life stages of organisms, but not all. Finally, some have questioned the original theoretical derivation of the ¾ law, arguing that the assumptions made by Brown, West, and Enquist are either incorrect

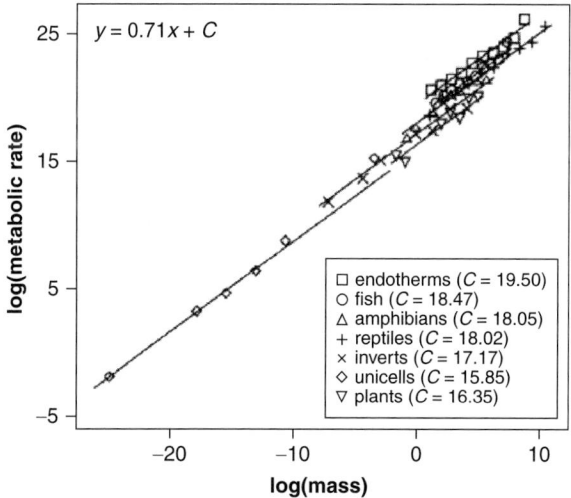

Figure 3.12 The log of metabolic rate plotted against the log of body mass for animals and plants. The slope of 0.71 is close to the value of 0.75 predicted by metabolic scaling theory. From Brown et al. (2004).

or unnecessary. A recent paper (Banavar et al., 2010) avoids some of the contentious assumptions of the original West et al. (1997) paper and yields new insights into the conditions under which ¾-power scaling for animals should arise.

The MST is not the focus of this book, and I do not intend to enter into the debate, except to offer a personal comment: every time I look at Figure 3.12 I am amazed by what it shows. In all of science, regularity across so many orders of magnitude is rare.

3.5 Why do we care about the metrics?

The empirical patterns we have been discussing appear to be pervasive across wide ranges of spatial scales, taxonomic groups, and habitats. That is evidence that what Darwin referred to as "the elaborately constructed forms" of nature have indeed "been produced by laws acting around us." And this is reason enough to care about macroecological patterns, for they are clues to understanding the nature of these laws.

But there are also more practical reasons to try to understand the patterns or forms these metrics take, not just pictorially from graphs, but if possible analytically. With analytical expressions, which means actual functions of independent variables, we can extrapolate information with greater confidence than we can from pictures. And as we shall see below, improving capacity to extrapolate knowledge brings us a big step forward in the effort to use ecological theory more unequivocally in conservation biology.

3.5.1 Estimating biodiversity in large areas

Estimates of the number of species of plants or birds, mammals or other classes of vertebrates at continental scale are fairly reliable, probably off by, at most, a few percent. But for insects, other invertebrates, and microorganisms, our knowledge of species' richness is much less complete. For example, intensive surveys of the canopies of individual trees in tropical forests always yield new species of arboreal beetles hitherto not known to science, so there clearly are more species of beetles than we currently catalog. We do not know, even to an order of magnitude, how many species of insects there are in the Amazonian rain forest, and uncertainty in our knowledge of microorganismal diversity is even greater. We care about how many such species there are because we would like to know what we are losing when habitat is lost, because it is inherently interesting to know what evolution is capable of producing, and because that knowledge helps us better understand evolution and ecology.

Consider the beetles in Amazonia. Only a small number of small patches of the vast forest can be censused directly, so how do we scale up knowledge from those patches to the entire biome? If one had confidence in an hypothesized functional form of the SAR across spatial scales that ranged from small patches to that of the entire biome, it would be easy. Just plug in the data at small scale and extrapolate the function describing the SAR to the desired large scale. But we can't have confidence in the hypothesized functional form of the SAR at large scale in the absence of data at that scale, which is just what we were trying to predict!

Clearly the only way out of this conundrum is to be able to test the hypothesized scaling law for species' richness all the way from small to large scales on at least some dataset or sets. Fortunately, such sets exist. Species' richness of plants, and of each class of vertebrates, are generally well estimated over continental and biome scales. One dataset that we will make use of in Part IV to test scaling theory is the product of extensive biodiversity surveys in the Western Ghats biome of southern India. There, for example, tree species' richness is well-estimated over 60,000 km^2, as well as on numerous much smaller plots scattered throughout the biome (Krishnamani et al., 2004).

But insects, other invertebrates, and microorganisms are not extensively censused, anywhere, at very large scales. If we test macroecological theory for plants or classes of vertebrates from small plots up to biome scale and it works, can we have confidence it will work for insects or microorganisms? We cannot answer that satisfactorily, but our confidence in doing so would surely increase if the same macroecological scaling laws that worked for higher organisms worked for these creatures from quite small scales (perhaps ¼ m^2) up to manageably larger scales such as 64 m^2.

The predicted number of species at very large spatial scales is quite sensitive to the form of the SAR used to extrapolate from data on small plots. Take the case of tree species in the Western Ghats. On 48 one-quarter ha plots, average tree species' richness is 32.5 species. Using the widely assumed power-law form of the SAR, with $z = ¼$, we can scale up from the 2500 m^2 area of a plot to the full 60,000 km^2 = 6×10^{10} m^2 area of the biome. This extrapolation yields $\bar{S}(60,000 \text{ km}^2) = 32.5 \cdot (6 \cdot 10^{10}/2500)^{1/4} \sim 2275$ species. This estimate is certainly too big. Extensive surveys of the Ghats have yielded to date no more than about 1000 species of trees (of greater than 10-cm circumference at breast height), and while a few new ones are being discovered each year, it is highly unlikely that the number will come close to our estimate.

One might argue that the power-law model might be fine but the assumption of $z = ¼$ is wrong. So suppose we determine the best power-law exponent by fitting the actual data from within the ¼ ha plots to a power-law model. Krishnamani et al. (2004) showed that that procedure yields an exponent of $z \sim 0.45$, which then predicts 68,000 species in the entire biome, an even worse estimate. In Chapter 8 we will show that METE actually makes a very realistic prediction for biome-scale tree diversity in the Western Ghats starting only with census data on the ¼-ha plots.

3.5.2 Estimating extinction

When a patch of the Amazonian rain forest is clearcut, some species may be driven extinct. If someday the only sizeable remaining forested portion of that forest is, let's say, just a 100 km × 100 km patch, even more species will be gone as a result. And if unmitigated global warming degrades the climatic suitability of the current habitats of plants and animals, we suspect that many species will also be lost. To estimate the number of extinctions in each of these situations, knowledge of our metrics is essential (May et al., 1995).

Consider the first case, in which a square patch of area 10^4 km^2 is cut out of the Amazon and, maybe pessimistically, assume that every creature in the eliminated patch disappears as a result. Of all the species found in the totality of the Amazonian Basin, those that will no longer exist, at least not in the Amazon, are the ones that were only found in that patch. They are exactly what we defined as endemics at 10^4 km^2 scale in Section 3.3.9: $\bar{E}(10^4$ km$^2)$. Actually the number we defined to be endemic at that scale is the average of the number of species unique to the cells of that scale, but if we assume that our clearcut cell is representative of all the cells, then the average is a reasonable estimate for the particular patch. So the EAR is the metric we use to estimate species' loss when a patch of habitat is removed (Kinzig and Harte, 2000).

In the second case a single square patch of area 10^4 km^2 remains after the Amazon is nearly all destroyed. As a first approximation (but see below) we can estimate the number of species remaining in that patch to be just the number that it sustained before the destruction. That would be $\bar{S}(10^4$ km$^2)$.

In the general case in which an area A_{before} is diminished to an area A_{after} the formula for the fraction of species lost, f, is:

$$f = [\bar{S}(A_{\text{before}}) - \bar{S}(A_{\text{after}})]/\bar{S}(A_{\text{before}}), \tag{3.27}$$

so if one had reason to trust a power-law form for the SAR:

$$f = 1 - (A_{\text{before}}/A_{\text{after}})^z. \tag{3.28}$$

Why did we use the EAR in the first case and the SAR in the second? In the first case, the area left behind after removal of the patch has a hole in it. It is not a nicely shaped area in the sense discussed in Section 3.4.7. So applying the SAR to the area of the intact Amazon minus the area that is clearcut might not give the correct answer.

The third case, species' loss under climate change, presents the most difficult conceptual issues. For one thing, nested and island SARs are applicable to different types of climate change scenarios. Consider first, the case in which a large biome, such as Amazonia, of area A_{before}, containing S_{before} species, shrinks under climate change to an area of A_{after} located within A_{before}. If the species are plants with limited dispersal capability, for example, how many species would we expect to find remaining in a randomly selected region of area A_{after}? The better answer is clearly given by the nested SAR. On the other hand, consider the case in which the suitable patch A_{after} is located completely outside the original A_{before}, and all the individuals remaining following climate change must disperse there from A_{before}. If the assumption is made that dispersal ability is not limiting and that individuals colonizing the

A_{after} are emigrating from the area A_{before}, then an island SAR form might be more appropriate. But dispersability is likely to be a limiting factor for some species, and thus knowledge of the metric $\Delta(D|n_0, A_0, \ldots))$ enters the picture.

In practice, the nested SAR and the island SAR (constructed as a collector's curve as described in Figure 3.2) approaches do give different estimates of the number of species expected to remain in a shrunken habitat A_{after}. Figure 3.13 compares estimates of species remaining using both methods, using data from the 50-ha Barro Colorado Island tropical forest plot in Panama. The nested-SAR approach consistently predicts higher levels of species loss than the island-SAR approach.

In most cases, neither a pure island- nor nested-SAR approach will be appropriate. First, dispersal capability will be neither unlimited nor completely absent. Second, newly suitable habitat is not likely to be only a sub-plot of the original habitat nor only disjunct from it. Third, for the case of disjunct habitats, the individuals found in

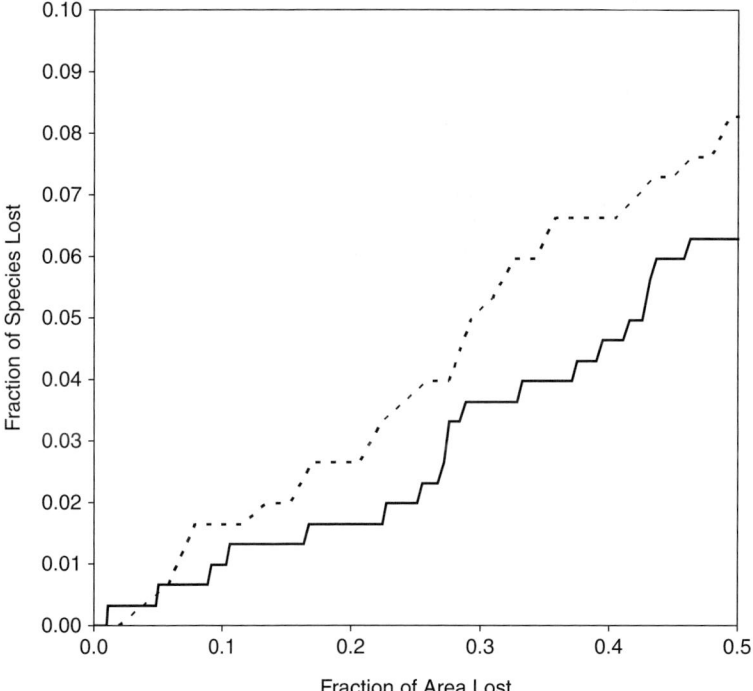

Figure 3.13 Differences between nested- and island-SAR approaches to estimating species remaining after habitat loss using data from the Barro Colorado Island 50-ha forest plot. The nested SAR (solid line) predicts more extinction after habitat loss than the island SAR (dashed line). The island SAR curve is constructed as a collector's curve (see text), averaged over 50 trials, calculated by sampling individuals without replacement from the total individuals' pool, assuming number of individuals is proportional to area (J. Kitzes, Pers.comm.).

A_{after} are not likely to be random draws from the species in A_{before}. As a result, it is impossible to choose a single SAR form, or exponent, to apply in all cases of habitat shifts under climate change.

Another knotty issue relevant to extinction from climate change, land-use change, or any other cause has to do with food web links. If all the prey species of a predator go extinct, then it is likely that so will the predator; or, if all the pollinators of a plant go extinct, then so will the plant. Thus knowledge of these potential "web cascades" is needed to fill out the picture of extinction. The metric $\Lambda(l|S_0, L_0)$ provides some of the information needed to infer likely extinction amplification due to web cascades, although only knowledge of the full web topology can give detailed and reliable predictions (Dunne, 2006; Srinivasan et al., 2007).

As a final example of the difficulty estimating extinction, and of the importance of our metrics, it should be understood that all of the applications of the SAR and EAR described so far apply to estimates of extinctions within a single habitat patch. We may more frequently, however, encounter landscapes in which the same collection of species is found in two or more locations, each of which shrinks in area. The problem then becomes that the SAR extinction calculations do not define which particular species are lost from each of the shrunken areas, and, as a result, the overlap in the list of lost species cannot be predicted. The number of species lost from a set of two shrinking patches could range anywhere from zero to the sum of the calculated losses in each patch individually. To help resolve this, knowledge of the species-level and community-level commonality measures, $C(A,D)$ and $X(A,D)$, are needed (Kinzig and Harte, 2000).

Summarizing this subsection, six metrics provide information needed to help us predict the loss of species under land-use and climate change: $\bar{S}(A)$, $\bar{E}(A)$, $C(A,D)$, $X(A,D)$, $\Lambda(l)$, $\Delta(D)$.

3.5.3 Estimating abundance from sparse data

Knowledge of the abundance of a species or of a distinct population within a species is critical to evaluating its conservation status (He and Gaston, 2000; Kunin et al., 2000). The rest of this subsection focuses on species, but the arguments apply to distinct populations as well. When its abundance diminishes, a species becomes more susceptible to extinction because of three phenomena. First, low abundance places the species at greater jeopardy of being wiped out either by fluctuations in death rates, say from environmental stress or disease, or by fluctuations in birth rates due to reproductive stochasticity. Second, a small population is more likely to lead to inbreeding and the impairments to fitness that are a consequence. The third phenomenon is related to the first, but is conceptually different as it applies with either deterministic or stochastic demographics; this is the Allee effect (Allee et al., 1949), which is a reduced probability that males and females can encounter each other when their population density is low.

The considerations above lead to the concept of a minimum viable population (Gilpin and Soule, 1986), which is one of the indicators conservation biologists use to assess the propensity of a species to go extinct. Thus, knowledge of the abundance of a species can help determine whether to assign it with the status of an endangered species. Unfortunately, however, abundance data are often not available, even for censused species, because many censuses only record the presence or absence of a species, on some array of plots or locations along transects, but do not attempt to count individuals. So suppose we have a large, gridded landscape that we believe to constitute the suitable habitat for some species, and on some fraction of the cells, each of a specified area, we have presence–absence data for the species. In other words, on that subset of cells we are given the information that the species is found to be present on a fraction f of them. Can we use our metrics to estimate abundance?

The answer is yes, with the estimate improved to the extent that (1) the censused cells are randomly located within the entire habitat, (2) the fraction of total habitat area covered by censused cells is large, and (3) that absence data can be trusted. This last point needs emphasis; if a cell is recorded to be the site of an observation of the species, that information can generally be trusted, but the non-sighting of an elusive species may or may not indicate it is truly absent.

With those caveats understood, the estimate of the abundance, n_0, can be obtained from our species-level spatial abundance distribution $\Pi(n|A, n_0, A_0)$. In particular we can express the fraction, f, of occupied cells as:

$$f = 1 - \Pi(0|A, n_0, A_0), \tag{3.29}$$

where A is the area of a censused cell, A_0 is the total area of habitat. If theory can provide the form of the dependence of Π on its conditional variables, then Eq. 3.29 readily leads to an estimate of n_0 from the measured estimate of the fraction f.

In the next chapter we turn to theories that have been previously proposed to describe, explain, or predict the forms of the metrics discussed here.

3.6 Exercises

Exercise 3.1

Pick one of the questions in Section 3.1.2 and outline a procedure that you would deploy to answer it. Describe each step, from walking out into the field, to the counting procedures, to arranging the data in a spread sheet, to the analysis and graphing of the data.

Exercise 3.2

Show that the discrete power-law distribution $P(n) = cn^{-a}$ corresponds to a rank–abundance graph that is a straight line on log(abundance) versus log(rank) axes, with slope $= (1-a)^{-1}$, and that the exponential distribution $P(n) = ce^{-an}$ corresponds to a straight line on an abundance versus log(rank) graph.

Exercise 3.3

Think about a collector's curve sampled with and without replacement. Which do you think will give a more rapidly rising curve?

Exercise 3.4

For the complete nested design, explain why the assumption that abundance is proportional to area is exactly correct.

Exercise 3.5

Explain why the nested species–area curve cannot increase faster than linearly in area; that is, if $S(A)$ is written in the form $S = cA^z$, then $z \leq 1$.

Exercise 3.6

This exercise will elucidate an important contrast between the EAR and the SAR. Look at your result for Exercise 3.3 and then explain why the complete nested endemics–area curve must increase at least as fast as linearly in area; that is, if $E(A)$ is written in the form $E = cA^y$, then $y \geq 1$.

Exercise 3.7

Examine all the metrics in Table 3.1 and figure out how many of them are truly independent, taking into account the interconnections among them implied by their definitions. Some of these interconnections are made explicit in the text, but be careful because there could be other interconnections not made explicit above.

Exercise 3.8

Using the actual values of S_0 and N_0 for the Serpentine data, show that our fitted function for the abundance distribution is properly normalized.

Exercise 3.9

Consider the made-up landscape that has been censused in a 4 × 4 grid, with results shown in the Figure 3.14. Each appearance of a letter is an individual, with the letter designating the species. For this landscape, graph the complete nested SAR and EAR at areas $A = 1, 2, 4, 8, 16$, where $A = 1$ corresponds to the single grid cell scale in the figure and at scales $A = 2$ and 8, all the non-overlapping 1×2 and 2×4 rectangles oriented in either direction are included in the analysis.

Exercise 3.10

For the landscape above, graph the rank abundance distribution for the entire plot with whatever choice of log or linear variables you think is most illuminating.

Exercise 3.11

Again for this made-up landscape, graph the species-level spatial abundance distribution at $A = 1$ scale for each of species a, b, j, and s.

aaa b ccccccc	b ccc e	aa c e ff g h	a bbb cc d g h j
a b cc e gg h nn	h jjj k n pp	g r	d gggg j mm q
ccc f k	j m	n pp	cc m s
b c h jjj k	c r ss	s u vv	s t u v

Figure 3.14 A spatially-explicit, fictitious community of species. Each letter is a single individual of the species labeled by that letter.

Exercise 3.12

To understand better the triphasic SAR, construct and graph on log–log axes the complete nested SAR for a landscape consisting of two adjacent habitats with distinct species. For simplicity, assume each of the two habitats is occupied by species in a pattern exactly as in the diagram accompanying Exercise 3.9, except that the species in the right-hand habitat will be labeled a',...,v', with the primes distinguishing them from the species in the left-hand habitat. Do this two different ways. First, construct your SAR at scales $A = 1, 2, 4, 8, 16, 32$, just as you did in Exercise 3.8, taking your rectangles oriented in both directions, but not including squares or rectangles that overlap the boundary between the two habitats. Then, for the same areas, include in your averages at each scale all possible squares or 1×2 shaped rectangles, including squares or rectangles for $A = 2, 4, 8, 16, 32$ that overlap the boundary between the two habitats. For example, for $A = 16$, there will now be a total of five 4×4 squares that go into the calculated average of $\bar{S}(16)$. How does the shape of the SAR depend on which way you do the averages?

Exercise 3.13

Consider the island SAR for real islands, with islands of differing area being colonized by a number of mainland individuals that is proportional to island area. If the resulting island SAR is conceptualized as being a collector's curve from the mainland individuals, discuss whether it should be thought of as a collector's curve with replacement, or without.

Exercise 3.14

Use the serpentine data (see Appendix A) to draw graphs of $\bar{E}(A)$ and $\bar{S}(A)$, and, for the three species for which $\Pi(n)$ was graphed in Figure 3.6, graphs of $B(A)$.

4

Overview of macroecological models and theories

Here you will get some exposure to the types of models and theories that have been advanced to explain the observed behaviors of the macroecological metrics. I have not attempted to provide a comprehensive and in-depth review of all the most widely cited current and past theories of macroecology. Rather, my goal is to introduce you to the essential concepts and some of the quantitative tools needed to understand a spectrum of theoretical frameworks. These frameworks exemplify differing approaches to making sense of the various patterns described in the previous chapter.

4.1 Purely statistical models

Recalling the discussion in Chapter 1, a statistical model entails drawing randomly from some hypothesized distribution and using the result of that draw to answer some question. Here we will review several statistical models that have been applied to macroecology.

4.1.1 The Coleman model: Distinguishable individuals

In a landmark pater, Coleman (1981) worked out in detail the consequences of assuming random placement of individual organisms in an ecosystem. The random placement model (RPM) predicts the species-level spatial abundance distributions, $\Pi(n)$, the box-counting area-of-occupancy relationship, $B(A|n_0, A_0)$, species-level commonality, $C(A, D)$, and the O-ring measure of clustering, $\Omega(D)$. It does not predict the SAD, $\Phi(n)$, however, so to use Eq. 3.13 to derive a species–area relationship in this model, an abundance distribution has to either be assumed or taken from observation. Coleman did the former, and predicted SAR shapes under various assumptions about $\Phi(n)$. The RPM also does not predict the form of any of our energy or mass distributions and relationships.

In the RPM, the probability that a species will have n individuals in a cell of area A if the species has n_0 individuals in a larger area, A_0, that includes A, is the binomial distribution:

$$\Pi(n|A, n_0, A_0) = \frac{n_0!}{n!(n_0-n)!} p^n (1-p)^{n_0-n}. \tag{4.1}$$

The probability, p, is just the probability that on a single random placement of an individual on to the plot A_0, the individual will land in the cell A. That probability is the fraction of the total area that the cell occupies, which means:

$$p = A/A_0. \tag{4.2}$$

Equation 4.1 can be understood as follows. The term $p^n(1-p)^{n_0-n}$ is the probability that n individuals are assigned to the cell, and $n_0 - n$ individuals end up outside the cell in a space of area $A_0 - A$. Importantly, this implicitly assumes that the probability any given individual is found in a particular cell of area A is independent of how many individuals are already found there or elsewhere.

The factorials in Eq. 4.1 arise because the n individuals that end up being found in the cell A can be any n out of the total number, n_0. For example, if $n_0 = 5$, $n = 2$, and if the individuals in the species are labeled a,b,c,d,e, then the two that end up in the cell can be any of the following pairs:

a,b; a,c; a,d; a,e; b,c; b,d; b,e; c,d; c,e; d,e.

There are 10 such pairs, which is 5!/2!3!. The symbol $\binom{n_0}{n}$, or the words "n_0, choose n," are often used to designate the combination of factorials in Eq. 4.1.

By assuming the combinatorics above, we are in effect assuming that the individuals within a species can be treated as distinguishable objects and that the order in which the individuals are assembled does not matter. If order mattered, then we would have to distinguish a,b from b,a, and we would be dealing with a different probability distribution. The case in which the individuals within each species are assumed to be indistinguishable and the case in which order of selection matters will be treated later in this chapter (Section 4.1.2).

The binomial distribution is sharply peaked about its mean, if the mean value of occupancies in the cells of area A is large compared to 1: $n_0 A/A_0 \gg 1$. In fact, it can be shown that in the vicinity of the mean value of n, the distribution approaches a Gaussian (Exercise 4.1a). On the other hand, when $n_0 A/A_0 < 1$, the distribution is monotonically decreasing. This behavior can be seen in the graphs in Figure 3.6, which show the RPM results along with data. When $p = A/A_0 \ll 1$, it can be shown that the binomial distribution is well approximated by the Poisson distribution (Exercise 4.1b).

Given the generally poor agreement between the observed spatial abundance distributions and the RPM predictions, as in Figure 3.6, one might expect from Eq. 3.13 that the RPM would not predict SARs very well. To use the RPM to predict the SAR, the total number of species in the plot and the distribution of their abundances $\{n_0\}$ have to be taken either from theory or from observation. Coleman (1981) and many publications since then (see, for example, Green et al., 2003;

Plotkin et al., 2000) have shown that few empirical SARs are well-described by the model.

The RPM predicts that species-level commonality, $C(A, D|n_0, A_0)$, will be independent of distance, D. This can be understood from the fact that two individuals randomly dropped on a plot are no more likely to land in two cells of area A that are close to one another than they are to land in two such cells that are far apart. The dependence of C on area, A, and inter-plot distance, D, can be worked out from Eq. 4.1, and we leave doing that as an exercise for the reader (Exercise 4.3).

The O-ring measure of clustering (Section 3.4.3) is predicted by the RPM to be $\Omega(D|n_0, A_0) = 1$. This is a trivial consequence of the way $\Omega(D|n_0, A_0)$ is defined as a ratio in which clustering is measured relative to random placement. Finally, the RPM prediction for $B(A|n_0, A_0)$ is immediately given by combining Eqs 3.7 and 4.1.

The failure of the RPM can be summarized as follows:

The observed values for $\Pi(0)$, the probability a species is absent from a cell, for most species at most spatial scales, but particularly for $A \ A_0$, exceed the RPM prediction. This smaller-than-observed prediction for the probability of absence implies that the RPM scatters individuals too uniformly over the plot, creating less aggregation than observed.

If the observed SAD is used in Eq. 3.13, and the RPM for $\Pi(n)$ is assumed, then observed SARs tend to increase more rapidly than predicted. This is directly tied to the fact that the RPM under-predicts aggregation: if individuals within species are predicted to be spread out more uniformly than observed, then at small scales, cells will be predicted to be more likely to contain any given species than observed. All else equal, greater clumpiness of the distribution of individuals within species produces steeper SARs.

We will see in Section 4.1.2 that if we give up the assumption of distinguishability of individuals, then the revised RPM performs more realistically.

4.1.2 Models of Indistinguishable Individuals

A rising area of theory building in ecology is based upon the assumption that differences among biological entities can be ignored. One such assumption, namely that differences in species' traits can be ignored, is the starting assumption for Hubbell's neutral theory of ecology, which I referred to briefly in Chapter 1.3.4 and which will be discussed in more detail in Section 4.3.2. Another type of neutrality, or symmetry as some authors prefer to call it, is to assume that while species' differences will not be ignored, differences among individuals within species will be. This assumption is surely no more unreasonable than assuming species' trait differences can be ignored, and in biological terms could be considered to correspond to ignoring intra-specific genetic differences.

Clearly, if you were studying natural selection in a species, so that genetic differences among individuals were essential to the phenomenon of interest, you could not get away with the assumption that individuals within a species are indistinguishable. But let us keep an open mind about whether neutrality could lead to a satisfying theory of macroecology.

To proceed, it will help to visit ideas first developed by the mathematical physicist Pierre Simon Laplace. He posed the following problem in probability analysis: you have the usual urn with two kinds of balls, red ones and blue ones, but you have no evidence whatsoever as to how many there are of each. Initially, they might all be blue, or all red, or any mixture of the two types. Suppose you reach in and pull out a red one and do not replace it. What is your best inference of the probability that the second ball you extract will also be red? Laplace showed that the answer is 2/3. In general, if there are two outcomes, R and B, and your prior knowledge consists only of knowing n_R extracted balls were red and n_B came up blue, then the best estimate of the probability that the next ball you extract will be red is:

$$P(n_R + 1 | n_R, n_B) = (n_R + 1)/(n_R + n_B + 2). \tag{4.3}$$

This is called Laplace's "rule of succession." Exercise 4.5 gives you the opportunity to understand this better by deriving it.

We can apply this rule to the assembly of individuals from a species onto a landscape, imagining the assembly process as a sequence of arrivals of individuals. We start out by populating the two halves of a plot. Thinking of the left half as red and the right half as blue, the first individual to enter the plot will have a 50–50 chance of entering on the left half. But as subsequent individuals are added, the probability of the next one in the sequence of added individuals going to the left half is given by:

$$\text{probability(next individual to left half)} = (n_L + 1)/(n_L + n_R + 2), \tag{4.4}$$

where n_L is the number on the left and n_R is the number on the right just before the next individual is added. This rule can be shown (Exercise 4.4) to generate the flat distribution:

$$\Pi(n | \frac{A_0}{2}, n_0, A_0) = \frac{1}{n_0 + 1} \tag{4.5}$$

We can understand this by thinking of the n_0 individuals as indistinguishable. Suppose $n_0 = 5$. Here are the six possible ways the 5 indistinguishable individuals can be allocated to the two halves of the plot:

$$(0, 5), (1,4), (2,3), (3,2), (4,1), (5,0).$$

If each of these has equal weight, which would be the case if the objects are indistinguishable, then each must have probability $1/6 = 1/(n_0 + 1)$. This idea, that each possible outcome has equal probability, is sometimes referred to as the principle of indifference: when there is no a priori reason to give two outcomes different probabilities, their probabilities must be assumed to be identical. The principle of indifference was first clearly developed by Laplace and it is the basis for his rule of succession, as you will see when you do Exercise 4.5. If the individuals were distinguishable, then there would be more ways in which the end result (2,3) could be obtained than there are ways to end up with (0,5). In fact there are 5!/(2!3!) more such ways and now the a priori probabilities of those outcomes differ.

There is another way to understand the difference between Eq. 4.5 and Eq. 4.1. In the latter, in the assembly process, the probability of an individual entering the left side is independent of how many are already on the left and right sides; for that case, p is always ½ in Eq. 4.1. But in Eq. 4.4, the probability of the next event is dependent upon the previous outcome. The assumption of indistinguishability and the assumption of the Laplace rule for the dependence on prior outcome (Eq. 4.4) lead to the same outcome for the allocation of individuals to the two halves of a plot!

We can understand this from another angle by considering a coin-toss approach to allocating molecules between two halves of a room. This is an example of what is called a Bernoulli process for generating a binomial distribution. On each toss, there are 50–50 odds of the molecule ending up in, say, the left half, and with a large number of molecules, the outcome will be a nearly equal division of molecules between the two halves. It actually doesn't matter whether the molecules are considered to be distinguishable or indistinguishable—the important assumption is that the outcome of each toss is independent of the previous outcomes. Picture a bag of coins, each from a different country and each with a heads and a tails. When each is flipped it is put in the left or right side of the room, depending on whether it landed heads or tails. The outcome at the end will be the binomial distribution, even though each coin was unique.

Note that the assembly rule in Eq. 4.4 is a "the rich get richer" rule, in which, as more individuals pile up on one half, either left or right, the probability of the next individual going to that half increases. In contrast, the binomial distribution result is egalitarian.

Readers with a physics background may see a similarity between the RPM with indistinguishable objects and Bose–Einstein statistics of indistinguishable particles in quantum statistics. Let's examine the case of classical statistics of distinguishable objects in a little more depth. In the room you are in, the gas molecules are nearly exactly equally allocated between the left and right halves of the room. The binomial distribution, Eq. 4.1, accurately predicts that the probability of even as little as a 1% or greater difference between the numbers of molecules in the two halves is negligible (Exercise 4.2). The narrow peak at the 50–50 allocation for the binomial distribution results from the large number of air molecules in a room, but even with only 100 molecules in the room, the probability of there being exactly a 50–50 split is much larger than the probability of there being, say, 5 on the left, and 95 on the right. With indistinguishable molecules, Eq. 4.5 predicts that there is a good chance that you would suffocate if you sat for a while in one end of a room! In fact, however, that is probably not the case. Were the air molecules to become detectably unequally distributed, a pressure gradient would arise and this would extremely rapidly re-adjust the distribution to near equality in numbers. Equalizing pressure forces, not the binomial distribution, is the reason you are not suffocating right now.

In Chapter 7 we shall see that METE predicts Eq. 4.5. Chapter 8, where we confront theory with data, will provide evidence that Eq. 4.5 describes adequately the distributions of individuals within a large fraction of, but by no means all, species at plot scales. To the extent that Eq. 4.5 describes the spatial distributions of some species better than does the binomial distribution, we can conclude that pressure forces (for example, from competition between individuals) are relatively weak for those species.

4.1.2.1 Generalized Laplace model

In 4.1.2 we examined the distribution of individuals across the two halves of a plot, applying a statistical rule that assumes individuals are indistinguishable. The question now arises as to the patterns that can be expected to result at finer spatial scales under that same assumption. This leads us to two choices: a generalized Laplace model and the HEAP model (see next section). We define and explain these in turn.

Consider the allocation of individuals to the four quadrants of A_0, each with area $A = A_0/4$. Now we have several choices. One possibility is to generalize the Laplace rule in Eq. 4.4 to give:

$$P(\text{next individual to upper left quadrant} | n_{UL}, n_{LL}, n_{UR}, n_{LR}) = (n_{UL} + 1)/(n_{UL} + n_{LL} + n_{UR} + n_{LR} + 4), \tag{4.6}$$

where, for example, n_{LR} is number of individuals found in the lower-right quadrant prior to the entry of the next individual. Let's work out an example. Suppose $n_0 = 2$. To calculate the probability that both individuals end up in the upper-left quadrant, first consider the probability that the first does so:

$$P(\text{first individual to upper left}) = (0+1)/(0+4) = 1/4. \tag{4.7}$$

This makes sense. The first individual to enter the plot has an equal probability of entering each quadrant, and since there are four quadrants, the probability must be ¼. If we multiply that probability by the probability that at the next entry step the second individual also goes to the upper-left, we obtain our answer. From Eq. 4.6:

$$P(\text{second individual to upper left} | 1, 0, 0, 0) = (1+1)/(1+4) = 2/5. \tag{4.8}$$

Hence:

$$P(\text{both individuals in upper left}) = (1/4)(2/5) = 1/10. \tag{4.9}$$

Note that for $n_0 = 2$, the possible outcomes (with indistinguishable individuals), are:

2 0	0 0	0 2	0 0	1 0	0 1	1 1	0 0	1 0	0 1
0 0	2 0	0 0	0 2	1 0	0 1	0 0	1 1	0 1	1 0

If each is equally likely, by the principle of indifference, the probability of each is 1/10, in agreement with Eq. 4.9. Equation 4.6 can readily be generalized to finer scales. We will call the spatial abundance distribution, $\Pi(n)$ that results, the "Laplace model." Box 4.1 describes this model in more detail.

Box 4.1 The generalized Laplace model

For cells of area A, where an integer number of such cells comprise A_0, and for abundance, n_0, in A_0, we can work out the prediction of the Laplace Model for $\Pi(n|A, n_0, A_0)$ by just counting distinct outcomes. First we have to calculate the total number of distinct ways, g (n_0, M_0), in which n_0 indistinguishable individuals can be placed into M_0 cells ($M_0 = A_0/A$). We saw above that $g(2, 4) = 10$, and in general it can be shown that:

$$g(n_0, M_0) = \frac{(M_0 + n_0 - 1)!}{(n_0)!(M_0 - 1)!}. \quad (4.10)$$

First consider $\Pi(n_0|A, n_0, A_0)$. Only one of the $g(n_0, M_0)$ outcomes will correspond to having all n_0 individuals in a particular cell, and so $\Pi(n_0|A, n_0, A_0) = 1/g(n_0, M_0)$. Next, consider $\Pi(0|n_0, A, A_0)$. The number of outcomes in which no individuals appear in a particular cell must equal the number of outcomes in which all the individuals appear somewhere in $M_0 - 1$ cells, which is given by $g(n_0, M_0-1)$. Hence, in the Laplace model, $\Pi(0|A, n_0, A_0) = g(n_0, M_0-1)/g(n_0, M_0)$. By the same reasoning, $\Pi(n|A, n_0, A_0)$ is given by the fraction of outcomes in which $n_0 - n$ individuals can be placed into $M_0 - 1$ cells, or:

$$\Pi(n|A, n_0, A_0) = \frac{g(n_0 - n, M_0 - 1)}{g(n_0, M_0)}, \quad (4.11)$$

where the g's are given by Eq. 4.10.

4.1.2.2 HEAP

The generalized Laplace model described above is only one way to model the spatial distribution of indistinguishable individuals. Another option is to iterate Eq. 4.4 at successively finer scales. This gives rise to a model that we have previously called HEAP (Harte et al., 2005), where HEAP refers to Hypothesis of Equal Allocation Probabilities. HEAP predicts a spatial abundance distribution that agrees with Eq. 4.4 at the first bisection, $A = A_0/2$, but at finer scales it diverges from Eq. 4.6 and the generalization of Eq. 4.6 to finer scales as given by Eq. 4.11. The details are explained in Box 4.2.

Box 4.2   How HEAP works

To understand how Eq. 4.4 can be iterated to finer scales, consider the example of $n_0 = 5$, and $A = A_0/4$. Here is how the probability of 5 individuals in the upper-left quadrant is calculated in the model:

$$\Pi(5|\frac{A_0}{4},5,A_0) = \Pi(5|\frac{A_0}{2},5,A_0)\Pi(5|\frac{A_0}{4},5,\frac{A_0}{2}). \tag{4.12}$$

In words, this equation just says that (the probability of 5 individuals in the upper-left quadrant) equals (the probability that there are 5 individuals in the left-half) times (the probability that if there are 5 individuals in the left-half then there are 5 individuals in the upper-left quadrant).

From Eq. 4.4, each of the two terms on the right hand side is equal to $1/(n_0+1) = 1/6$, and so the answer is 1/36.

What about the probability that there are 4 individuals in the upper left quadrant? This is now equal to the sum of two products:

Probability of 4 individuals in the upper left quadrant = [(probability of 5 individuals in left half) ×

(probability that if there are 5 individuals in the left half, then there are 4 individuals in the upper left half)]

+

[(probability of 4 individuals in left half) x
(robability that if there are 4 individuals in the left half, then there are 4 individuals in the upper left half)]. (4.13)

In equation form, this reads:

$$\Pi(4|A_0/4,5,A_0) = \Pi(5|A_0/2,5,A_0) \times \Pi(4|A_0/4,5,A_0/2) + \Pi(4|A_0/2,5,A_0)$$
$$\times \Pi(4|A_0/4,5,A_0/2) = (1/6)\times(1/6) + (1/6)\times(1/5) = 11/180. \tag{4.14}$$

More generally, the HEAP probabilities are calculated from the following recursive equation of which Eq. 4.14 is just an example:

$$\Pi(n|\frac{A_0}{2^i},n_0,A_0) = \sum_{q=n}^{n_0} \Pi(q|\frac{A_0}{2^{i-1}},n_0,A_0) \cdot \Pi(n|\frac{A_0}{2^i},q,\frac{A_0}{2^{i-1}}) =$$

$$\sum_{q=n}^{n_0} \frac{\Pi(q|\frac{A_0}{2^{i-1}},n_0,A_0)}{(q+1)}. \tag{4.15}$$

Box 4.2 (*Cont.*)

In words, this equation states that (the probability of n individuals in a small cell, A, of area $A_0/2^i$, given n_0 at largest scale, A_0) equals the sum over all q-values greater than or equal to n of (the probability that there are q individuals in a cell of area $2A$, given n_0 individuals in A_0) times (the probability that if there are q individuals in that $2A$ cell, then there are n in one half of that cell). By Eq. 4.4, that last probability is $1/(q+1)$.

In the HEAP model, the 10 diagrams shown above (Section 4.1.2.1) that arise when placing 2 indistinguishable individuals into 4 boxes do not all have equal probability. The actual probabilities for each of the diagrams can, in that example, be calculated from cross-scale consistency relationships however. We return in Chapter 11 to a discussion of cross-scale consistency relationships and their implications for predicting the distance-decay function discussed in Section 3.3.3

4.1.3 The negative binomial distribution

Yet another statistical model, the negative binomial distribution (NBD), was proposed as a good description of the spatial distribution of individuals within species by Bliss and Fisher (1953). In its original form and usual applications, it describes the distribution of an integer-valued variable over an infinite range: $n = 0, 1, 2, \ldots \infty$. For our spatial abundance distribution Π, the infinite-range NBD is:

$$\Pi(n|A, n_0, A_0) = \frac{(n+k-1)!}{n!(k-1)!} \left(\frac{\bar{n}}{k+\bar{n}}\right)^n \left(\frac{k}{k+\bar{n}}\right)^k. \quad (4.16)$$

Here $\bar{n} = n_0 A/A_0$ and k is an adjustable parameter that describes the level of aggregation in the spatial distribution. In the limit of $k \to \infty$, this distribution approaches the Poisson distribution $\bar{n}^n e^{-\bar{n}}/n!$ and for $k = 1$ the distribution is the exponential $\exp(-\lambda n)$ with $\lambda = \log(1+\bar{n}^{-1})$. In this model, the probability of a species being present in a cell of area A is given by $p = 1 - [(k+\bar{n})/k]^{-k}$ and the variance of the distribution of abundances within a cell is related to the mean cell abundance by $k = \bar{n}^2/(\sigma^2 - \bar{n})$.

Many authors (Wright, 1991; He and Gaston, 2000; Plotkin and Muller-Landau, 2002) have explored the applicability of this model to species-level spatial aggregation patterns, to SARs, and to species' turnover, as expressed by our commonality metric $C(A, D)$. The parameter k can be species- and scale-dependent, and thus the model provides considerable flexibility to describe different species over a range of spatial scales.

Referring to Figure 4.1a and b, we see that the negative binomial ($k = 1$) model and the Laplace model are in close agreement. We can now see why this is. For the negative binomial distribution, with $k = 1$, Eq. 4.16 results in:

$$\Pi(0|A, n_0, A_0) = \frac{A_0}{A_0 + n_0 A}. \quad (4.17)$$

96 • *Overview of macroecological models and theories*

(a)

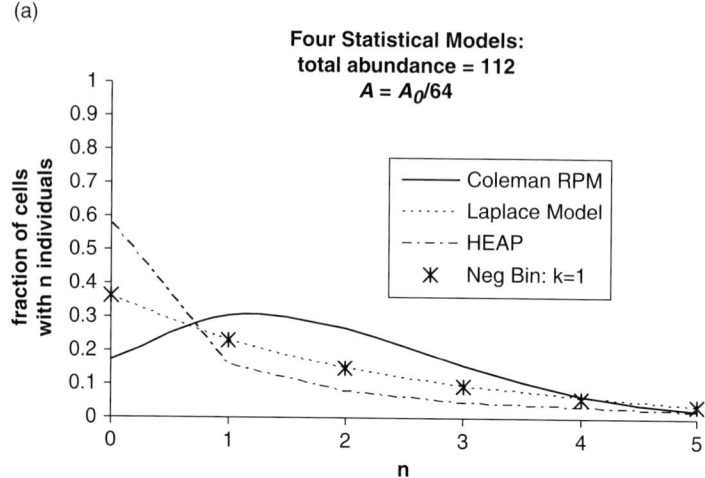

(b)

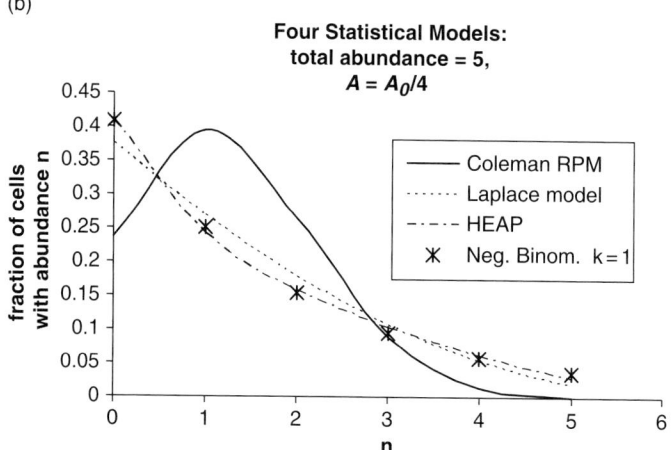

Figure 4.1 The predictions for $\Pi(n)$ from four spatial distribution models and two combinations of n_0 and A/A_0: (a) $n_0 = 112$, $A = A_0/64$; (b) $n_0 = 5$, $A = A_0/4$.

From Eqs 4.10, 4.11 (Box 4.1), we can derive the comparable Laplace model result and examine it in the limit $A_0/A \gg 1$:

$$\Pi(0|A, n_0, A_0) = \frac{(\frac{A_0}{A} + n_0 - 2)!(\frac{A_0}{A} - 1)!}{(\frac{A_0}{A} + n_0 - 1)!(\frac{A_0}{A} - 2)} = \frac{\frac{A_0}{A} - 1}{\frac{A_0}{A} + n_0 - 1} \simeq \frac{A_0}{A_0 + n_0 A}. \quad (4.18)$$

This agrees with the result from the negative binomial distribution in Eq. 4.17. It is left as an exercise (Exrcise 4.6) to show that this close agreement of the two distributions extends to larger values of n.

Figure 4.1 also shows agreement between the HEAP distribution and both the Laplace and negative binomial distributions at coarse scale ($A = A_0/4$), low abundance, but not at fine scale, higher abundance. It is scale that matters here, as you can see by recalling that at the scale of a single bisection, $A = A_0/2$, the Laplace and HEAP models are equivalent.

4.1.4 The Poisson cluster model

The Poisson cluster model (see, for example, Plotkin et al., 2000) has even more parameter flexibility than the negative binomial. It assumes that individuals within any given species form multiple aggregates (clusters) within a landscape and these aggregates are randomly (Poisson) distributed over the landscape. Assuming that within the clusters the individuals are densest at the center, falling off according to a Gaussian distribution toward the edge of the cluster, and picking reasonable parameters to describe the Gaussian, the model can adequately describe empirical spatial patterns, and greatly improves upon the RPM predictions. Whether the model is a more accurate predictor of spatial pattern than the NBD or the HEAP distribution or other predictions distributions is perhaps not as interesting as is understanding its origins, either in explicit ecological mechanisms or in some other theoretically based set of ideas, deriving the parameters that characterize the Gaussian distributions, and then deriving from those same mechanisms or ideas predictions for other metrics in Table 3.1. For example, because the aggregates are randomly distributed over the landscape, a prediction of the model is that species-level commonality, $C(A, D)$, should be independent of D once D exceeds the diameter of a typical cluster.

4.2 Power-law models

Finding or assuming power-law behavior in nature is closely tied to the concept of scale-invariance. Consider the following non-power function of a variable, x:

$$f(x) = e^{-\lambda x}, \qquad (4.19)$$

where λ is some constant. The product λx must be unit-less... it cannot have units of time or distance or mass or any other fundamental quantity. To see why, consider the function e^w, which can be written, identically, as:

$$e^w = \sum_{i=0}^{\infty} \frac{w^i}{i!}. \qquad (4.20)$$

(In this expression, recall that $0! \equiv 1$.) Because every integer power of w appears in this expression, it is clear that if w had units of, say, mass, then the function e^w would have mixed units... mass to every power. That makes no sense, and so w must be unit-less. Hence if x in Eq. 4.19 has units of, say, mass, then λ must have units of mass^{-1}. Now suppose we change the units in which we are measuring x, perhaps changing from grams to tons. A corresponding change in the units of λ is required, which means the constant, λ, in the function is altered. That, of course, is not a fundamental problem and exponential functions widely arise in science. But it does indicate that the function is scale-dependent: its numerical form depends upon the units used to describe the variable, x.

Contrast that with a power function: $f(x) = c \cdot x^\lambda$. Here, λ must be a unit-less number because otherwise, for the same reason as above, the function $f = c \cdot x^\lambda = c \cdot \exp[\lambda \cdot \log(x)]$ would have mixed units. Now let the units of x change so that $x \to x' = ax$. This results in a change in the function $f(x) \to f(x) = cx'^\lambda = c' \cdot x^\lambda$, where $c' = c \cdot a^\lambda$. So now the constant out in front changes in value to reflect the units change, but the numerical form of the power-law remains the same. In particular, the exponent, λ, has not changed. In that sense, a power-law behavior is said to be an indication of scale-invariance in the system being described.

To explain the concept of scale-invariance in ecology it helps to consider what happens as we change the level of detail at which we observe an ecosystem pattern. Suppose that from high up in a balloon you look down on a huge area of forest and make a rough drawing showing the distribution of forest cover. No individual trees can be resolved from that height, but there will probably be large areas where the forest looks continuous and other places where there are large clearings in the forest. Now lower the balloon until you are looking at perhaps just a square kilometer of forest. Now you still cannot resolve individual trees but here and there you see smaller forest gaps. Despite the huge difference in the scale at which you are looking, and the level of detail you can resolve, you could conceivably find that if you are only concerned with the clustering pattern in the greenness below, you actually cannot tell whether you are looking from far away and seeing a coarse pattern or looking from nearby and seeing finer detail in the pattern of open spaces. You might even descend further and just stare at a single canopy, noting where the leaves are clustered and where there are gaps in that canopy. Again you might find that the clustering pattern is the same as it was from higher up. Scale-invariance refers to that situation: the same patterns emerge no matter the scale at which you observe the system.

Sometimes, when ecologists photograph a plot on which they are working, they place a Swiss army knife on the ground to provide a sense of scale. Figuratively speaking, scale-invariance means that nature does not come with a Swiss knife. But clearly, scale-invariance cannot hold across all scales in the real world. Even if our forest really were scale-invariant over the range of scales mentioned above, eventually, if we go high enough, we might see the edge of the continent, while if we get up close enough, all we see is a single cell within a leaf. We know that at those very large and very small scales the patterns will be completely different, so we are able to know the scale at which we see them.

The terms scale-invariance, self-similarity, and fractal are often used interchangeably. More interesting than the differences in some authors' usages of the terms are the different types of patterns that the terms can be applied to. A pattern can be either statistically self-similar (or fractal or scale-invariant), or exactly self-similar (or fractal or scale-invariant). And the terms can apply to a pattern over an infinite scale range or only over a finite scale range. Statistical self-similarity is, in fact, the type of self-similarity that most scientists have in mind when they assert that a coastline or a forest or the branching network in the blood circulatory system is fractal. The self-similarity is only statistical because any particular segment of, say, a coastline or mammalian circulatory system, is probably unique in some details. At best it is only averaged properties of the segments and branching patterns that exhibit self-similarity. Moreover, self-similarity in nature is always bounded. Following along the circulatory system, at some point one comes down to terminal capillaries or up to the aorta, and there the self-similarity ends. Thus, self-similarity in nature is sometimes associated with what is called "finite-size scaling." Mathematics can generate exactly self-similar shapes over an infinite range of scales, but only because mathematics allows us to generate precisely defined shapes and to pass to the limit in which the scaling range extends down to the infinitesimal and up to the infinite.

Were power-laws a good description of species–area relationships, the implications could be profound. Suppose you have an ecosystem of indeterminate size and you want to know the form of the SAR within that ecosystem. Suppose, further, that the system contains no intrinsic scale length, by which I mean the following: other than the quantity A, there are no other quantities with units of area, or area to some other power than 1, that are intrinsic to, or describe, the system. For example, say you are somewhere in the middle of the Amazon looking at species' richness at different spatial scales within a 50-ha plot, and you do not think the sheer size of the whole Amazon, A_0, can possible influence the shape of your SAR; hence the quantity A_0 will not appear in an expression for the SAR. Moreover, you do not believe that within your plot there is any intrinsic distance scale governing the abundance distribution of the species, other than the size of the cell in which you are measuring species' richness. That means that there is no unit-less quantity λA that can be formed from a quantity λ with units of area^{-1}, to place in the exponent of the SAR to form $\overline{S}(A) = S(0)e^{\lambda A}$. So that form, and by the same argument anything other than a power-law form, can be ruled out.

In the case of a power-law $\overline{S}(A) = cA^z$, the constant c does have strange units; because S does not have units of area, c must have units of area^{-z} to balance the units of area on both sides of the equation. That is not a problem, because, as argued above, a change in units will not affect the form of the SAR. The numerical value of the constant c is not an intrinsic property of the system, whereas if the power-law model were correct, the value of z would be. For any specified units of area, the value of c is the number of species at $A = 1$ (in those units), and so values of c for different ecosystems can be usefully compared at $A = 1$ to provide a measure of relative species' richness only at that area in those units.

Now, if you really were in the middle of the Amazon, conducting a census of trees or insects or birds and constructing an SAR at scales much smaller than that of the entire Amazon, you would readily observe characteristic length or area scales in the system... that is, quantities with units of length or area that "matter" to the way in which individuals and species are distributed. For example, the size of an individual tree canopy might play a role in shaping the SAR for arboreal insects over a range spanning sub-canopy to multiple canopy areas. The size of a canopy, and also of the root system, might influence the SAR for trees. Other length or area scales can arise from the abiotic environment, which is characterized by many spatial scales over which topography, climate, and soil properties vary.

In other words, the notion of a scale-free system is not likely to be applicable to real ecosystems, and thus a potentially compelling argument for power-law behavior of the SAR, scale-free dynamics, is hard to defend. Nevertheless, power-law SARs are widely discussed in the literature, sometimes claimed to characterize data, and often assumed to be a useful approximation in extrapolations of the SAR in conservation biology (Section 3.5).

Deciding when to adopt a power-law model based on straight-line fits to log–log graphs is not easy. In Box 3.3 we discussed the general concept of a power-law SAR, but what sort of model would actually generate self-similar landscapes and a power-law behavior for that metric? Before answering that, let us look at the simpler concept of a self-similar model for the species-level range-occupancy relationship $B(A|n_0, A_0)$. Once you understand how that is done, you can then explore the implications for the SAR of that same model by combining Eqs 3.3 and either 3.13 or 3.14. Appendices B.1 to B.4 provide a detailed look at power-law models in macroecology, with an example of a way to generate power-law behavior for $B(A|n_0, A_0)$ the topic of B.1.

Using the concepts laid out in Appendix B.1, Appendix B.2 extends the model to the SAR, and, unexpectedly, shows that in general it is impossible to have both power-law $B(A)$ for all species and a power-law $\overline{S}(A)$. The species-level spatial abundance distributions $\Pi(n|A, n_0, A_0)$ can also be derived for a scale-invariant ecosystem, as shown in Appendix B.3. The details of the derivation of the commonality metric, $X(A,D)$, in a scale-invariant ecosystem are given in Appendix B.4; because that result has been applied by various authors to determine SARs at scales for which direct measurement of species' richness is not possible, I discuss the relationship between commonality and the SAR in the following section.

4.2.1 Commonality under self-similarity

The assumption that a system is self-similar or scale-free has interesting implications for the community- and species-level commonality metrics. Consider the community metric $X(A, D|...)$, and assume that the only two quantities with units of length or area that it depends upon are A and D themselves. This would be in contrast to the explicit assumption about contingency made in Table 3.1, where X was listed as being contingent on A_0, the size of the system. If X depends only on A and D, then it must be a unit-less function of the ratio A/D^2: $X = X(A/D^2)$. Without using further information, we can say nothing more specific. In fact, we cannot conclude, just from the above reasoning, that X is a power-law function of A/D^2. We would expect that commonality decreases with increasing distance between censused patches, and thus if the function does obey a power-law, then it goes like (A/D^2) to a positive, not a negative, power, but we can not conclude anything more than this with dimensional arguments alone.

Pursuing the idea of scale-free systems a little further, however, we show in Appendix B.4 that in a scale-free world, there is a connection between the commonality function and the species–area relationship. In particular, the dependence of X on A/D^2 would indeed be a power-law. The derivation is based on first holding D fixed and varying A. Using the result from Appendix B.2 that the species–area relationship under self-similarity is a power-law, B.4 shows that the A dependence of X must also be power-law, with the same z-value as in the SAR, when D is held fixed. But then, because X can only depend on the combination A/D^2, the functional form of X in a self-similar world must be:

$$X(A, D) \approx (A/D^2)^z, \qquad (4.21)$$

For the macroecologist seeking far-reaching insights into patterns in the distribution of species, it is unfortunate that the world is not self-similar, because Eq. 4.21, were it applicable, would offer much insight. One application of this equation is to the determination of the slope of the SAR at large spatial scales in situations where complete nested censusing is not possible. By fitting distance decay data (i.e. commonality at fixed A, increasing D), to Eq. 4.21, SAR slopes, z, at large spatial scales can be empirically determined.

That was the procedure used by Green et al. (2004) and Horner Devine et al. (2004) to estimate z-values for microbial populations in soils. Krishnamani et al. (2004) also used this procedure to estimate z-values at scales up to several thousand kilometers in the Western Ghats Presesrve of India, using data on distance decay between ¼-ha plots scattered throughout the Preserve. The application by Krishnamani et al. did not assume a constant z-value across all scales. They used nested census data at $<$ ¼-ha scale to obtain z at those small scales, and using Eq. 4.21, found z at large scale to be considerably smaller than at small scale.

The species-level commonality metric in Table 3.1a is also constrained in the self-similarity model. If self-similarity holds at the species-level for some species, we have shown using a scaling argument (Harte et al., 1999b) that the community-level commonality metric behaves as $C(A, D|n_0, A_0) \sim (A_i/D^2)^y$, where $y = -\log_2(\alpha)$ is the power-law exponent for the range-area relationship and α is defined in Appendix B.1.

4.3 Other theories of the SAD and/or the SAR

4.3.1 Preston's theory

Arguably the most widely cited and influential set of theoretical papers on the SAR and its relationship to the SAD are those of Preston (1948, 1962) and May (1975). Their work concluded that the lognormal distribution provides a reasonable description of species' abundance distributions and that the species–area relationship is to a good approximation a power-law with exponent $z \sim \frac{1}{4}$. I urge readers to go back and read the original papers of these authors for they are classics. Here I will present some general comments about lognormality in ecology, and then summarize and critique the strategy underlying the Preston and May work.

Arguments that SADs are lognormal are sometimes based on the central limit theorem (CLT). In its most general form, the central limit theorem (CLT) informs us that if we repeatedly randomly sample a set of numbers from a larger set of numbers, then the distribution of averages of the sets of numbers approaches a normal distribution. Similarly, the distribution of the geometric averages of the sets of numbers approaches a lognormal distribution. Another commonly encountered statement of this theorem is that if a variable is the sum or average of a set of independent and randomly-distributed variables, then repeated estimates of that variable approach a normal distribution. Related to this second formulation, if a variable is the product, or geometric average, of a set of independent and randomly-distributed variables, then repeated estimates of that variable approach a lognormal distribution.

This last formulation of the CLT has been asserted to be relevant to the SAD. The underlying assumption is that the population size of a species is governed by many factors, such as light, a variety of nutrients, and water for plants. Further, these factors arguably combine as a product, not a sum, because if any of those factors are absent, population growth cannot occur. If this is true, and the availability of these resources is random, then a lognormal distribution is imaginable. Several problems arise, however. First, it is individual growth rates that are more directly related to the product of growth-enhancing factors, not population size, and normally distributed growth rates are unlikely to lead to normally distributed population sizes. Similarly, lognormally distributed growth rates are unlikely to lead to lognormally distributed population sizes. Second, the factors that influence population growth are nearly certainly not all independent; for example, high light levels might induce more growth, which could decrease nutrient supply, or less water might reduce nitrogen turnover, decreasing the amount of plant-accessible nitrogen. Because the factors contributing to growth are not independent, one must fall back on the first formulation of the CLT above. But the relevance of this to the distribution of population sizes across species is unclear; again, the distribution of growth rates across individuals might conceivably be lognormal by this argument, but why should it be relevant to population sizes across species?

Empirically, few abundance datasets exhibit the symmetry of a lognormal when the SAD is plotted as fraction of species versus binned logarithmic intervals. In particular, empirical SADs often exhibit asymmetric shapes, with fewer very abundant species and more rare species than would be predicted by the lognormal. One might argue that this is a sampling problem, and that those rare species would be seen to possess greater abundance if the sample size increased by more thorough sampling. In other words, a greater sampling effort might reveal a smaller fraction of truly rare species and thus a more symmetric distribution (plotted against log (abundance). On the other hand a greater sampling effort might just reveal more rare species. Williamson and Gaston (2005) and Nekola et al. (2008) summarize a number of conceptual and empirical problems with the assumption that the SAD is lognormal, and also provide a critique of a form of the sampling argument that is sometimes referred to as the Preston veil argument (Preston 1962).

Putting aside these concerns, let us continue with Preston's theory. The lognormal SAD contains two parameters: the mean and variance of the logarithm of abundance. Implicitly two other parameters play a role: the total abundance and the total number of species in the system. Preston used the following argument to conclude that there was a relationship among them. By dividing the abundances into "octaves," with each octave differing in abundance by a factor of two compared to an adjacent octave, a plot with octave number on the x-axis provides one way of logarithmically displaying abundance. On that axis, two functions can be plotted. The first is the fraction of species with abundance in each octave; this is just the usual SAD. The second is the total number of individuals in each octave, which gives a shape called the "individuals' curve." Preston (1948) had noted that in real datasets there is a tendency for the peak of the individuals' curve to occur at the highest octave represented in the SAD—that is, at the octave containing the most abundant species. Preston called such lognormal abundance distributions "canonical" and posited their prevalence in nature.

Continuing, Preston showed that sampling from a canonical lognormal abundance distribution generates a species' list that grows as a power of the number, N, of individuals sampled. He obtained a value of ~ 0.26 for the exponent of this power-law. To make the connection between the number N and the area of sampled ecosystem, Preston made the reasonable assumption that N should be proportional to area sampled. That then leads to an SAR of the form $S(A) \sim A^{0.26}$, a remarkable result because a power-law with $z = ¼$ has been widely thought to be the "true" form of the SAR.

There are fundamental problems here. Preston's argument is really about a collector's curve, not a nested species–area curve. In fact, referring to Eq. 3.13, the abundance distribution alone cannot possibly pin down the shape of the SAR; that shape must depend upon the spatial distribution of the individuals within each species. If the Preston result applies somewhere it would be to island SARs, where the individuals on islands can be thought of as representing the equivalent of a collector's curve of mainland individuals. In other words, if individuals from the mainland randomly fly out and settle on islands in numbers proportional to island area, and if the mainland SAD were the canonical lognormal, then the island SAR would be, by Preston's argument, $\bar{S}(A) \sim A^{0.26}$. But the evidence for such a power-law SAR was thought to come mainly from nested SARs. Indeed, as we showed in Section 3.4.7, collector's curves are generally better fit by logarithmic functions: $\bar{S}(N) \sim \log(N)$ than by power-laws.

Thus, for multiple reasons (validity of lognormal SAD, confusion between collector's curves and nested SARs, applicability of collector's curve to island occupancies, inadequate fit to actual collector's curves, validity of the sampling argument) the status of this theory remains in doubt. Some additional problems with the original Preston derivation were addressed by May (1975), but his fundamental finding was in overall agreement with Preston's. The Preston–May work is brilliant in execution, but of doubtful relevance to nested SARs.

4.3.2 Hubbell's neutral theory of ecology

The original NTE of Hubbell (2001), with later refinements by many other researchers (e.g. Volkov et al., 2003, 2004; Chave, 2004; Etienne et al., 2007) provides a set of models of the time dependence of species' abundance and composition in an ecosystem in which birth and death events occur stochastically. The term neutral means that the distributions from which the birth and death rates are drawn are assumed to be the same for all species. In fact, in neutral theories, neither species, nor individuals within species, have any traits that distinguish them from one another a priori. Distinctions arise only from the outcome of stochastic demographics.

In the theory, species can go extinct, and new species can be formed by speciation, with the speciation rate also assumed to be independent of traits. When an individual undergoes a stochastic death, a new individual born to an existing species or of a newly formed species, is assumed to take its place. In a variant on the Hubbell theory, Bell (2001) replaces speciation with migration to the plot of new species from outside the plot. The Hubbell theory has several fundamental tunable parameters: a speciation rate, a measure of the size of the community (a carrying capacity), and what is effectively a time constant representing an inverse birth rate. The NTE predicts two SADs. One is for the metacommunity; this is the large collection of species and individuals that provide the pool of entrants into the plot being modeled, and the other SAD is for any particular plot that receives immigrants from the metacommunity. The metacommunity SAD is predicted to be the logseries distribution (see Box 3.3) and the predicted plot SAD is called the zero-sum multinomial distribution. It describes quite well the SAD observed in the 50-ha BCI tropical forest plot (Hubbell, 2001; Volkov et al., 2003).

The mathematical formalism used by Hubbell and others to extract an SAD from theory is somewhat complicated, deploying what is called a "master equation" to look at the sequential consequences of birth and death events, as well as migration or speciation. Conlisk et al. (2009) simplified the mathematics considerably by showing that a simple assembly process generates exactly the same plot-scale SAD as does the master equation for the local community.

With the SAD in hand, a collector's curve can be calculated, and, following Preston, if N is assumed to be proportional to area, a kind of SAR, perhaps applicable to islands, is obtained. The theory as described above can only predict the shape of the nested SAR if spatial distributions of individuals within species, the $\Pi(n|A, S_0, N_0, A_0)$, are predicted. To accomplish this a dispersal distribution, $\Delta(D)$, can be introduced. In keeping with the neutrality concept, the shape of this distribution is assumed to be identical for all species, and thus independent of n_0. Hubbell (2001), and also Durrett and Levin (1996), examined nearest-neighbor dispersal. Chave and Leigh (2002), Etienne and Olff (2004a, 2004b), Zillio et al. (2005), and others have run the NTE assuming a Gaussian function for the dispersal distribution $\Delta(D)$, and Rosindell and Cornell (2007), using a wider variety of dispersal distribution, argued that restricting dispersal to sites nearest parents is biologically unrealistic and, from their investigation of more general dispersal distributions, concluded that the dependence of the resulting SAR on the dispersal distribution is relatively weak.

Interestingly, Rosindell and Cornell show that appropriate dispersal kernels can result in the triphasic form of the SAR (Section 3.4.7), in which, as area increases, species' richness increases steeply at small area (so-called local scale), more shallowly at intermediate area (regional scale), and then steeply again at larger areas (continental scale). The Rosindell and Cornell study is noteworthy because it does not need to assume wrap-around boundary conditions (equating the far edges of the plot with each other) when simulating landscapes, but rather passes to the limit of an infinite landscape. This is not only important because wrap-around condition can introduce artificial patterns, but also because the break-points separating local from regional or regional from continental triphasic behavior of the SAR in finite systems appears to depend on the assumed size of a finite system. In all such models the actual slopes of the SAR depend upon the fundamental parameters in the theory.

The literature on the NTE is vast and varied. The theory has its critics (see particularly Clark (2009) and Purves and Turnbull (2010), who rightly remind us that species really do differ in their traits, and thus should be expected to differ in their demographics in ways that extend beyond stochastic variations in the outcome of demographic events. Other critics have argued that the empirical evidence in support of the NTE is not compelling. Nevertheless, it is difficult to react with anything other than amazement that a theory based on such simple and counterfactual assumptions is as successful as it is.

4.3.3 Niche-based models

Yet another class of models has it origin in the observation that the growth of organisms and the persistence of species depends upon the ability of individuals to find and extract suitable energy and material resources from their biotic and abiotic environment, to avoid being eaten, to reproduce, and to adapt to changing conditions. The characteristic properties of organisms that allow that to happen are called traits. The number of possible traits is immense; just to list a few, animal traits include coloration, speed, olfactory ability, toxicity, thermal tolerance, gestation period, and gape size, while plant traits include rooting depth, possession of antifreeze chemicals, ability to form symbiotic associations with microorganisms, leaf thickness, lignin:nitrogen ratio of foliage and pattern of leaf venation. And the number of possible values that particular traits can have is uncountable. The niche a species occupies can be thought of as the collection of trait values possessed by the individuals in the species that allows the species to persist in its habitat.

Niche-based theories of the SAD proceed by the following line of reasoning. In any ecosystem, competition, predation, disease, and the vagaries of the abiotic environment, may result in the extinction of some of the species. If two species occupy nearly the same niche, meaning that they eat and are eaten by the same species, they capture food and resist predation with the same capacities, and their abiotic needs, such as preferred temperature conditions, are similar, then competition between them is likely to eliminate one (Gause, 1934). Thus, the reasoning goes, persistent species should differ in the combinations of trait values that they possess. The space of all possible values of traits ("niche space") can be thought of as being partitioned somehow across species. Hutchinson (1957) envisioned niche space as a very high dimensional coordinate system, with each axis representing some resource or other condition in the environment that influences the growth, survival and reproduction of individuals. For each species, with its range of values of traits, the amount of resources the species can extract and use as a consequence of those trait values, will influence its survival and reproductive success. The rules for apportioning trait space and relating trait values to growth, predator avoidance, reproductive success, and hence to abundance, constitute niche-based theory of the SAD.

The literature based around the formulation of such rules is enormous. Tokeshi (1993) provides an excellent review. One famous and influential example of a niche-apportionment rule is the broken stick model of MacArthur (1957, 1960) and later refined by Pielou (1975), Sugihara (1980), Tokeshi (1993), and Nee et al. (1991). Imagine taking Hutchinson's high-dimensional niche space and collapsing it on to a line that stretches from 0 to some finite positive number. An interval along the line represents a share of all the components of niche space available to a particular species. A broken stick model consists of particular rules for dividing the line into segments. Abundance distributions result if the assumption is made that longer pieces of the niche line correspond to more resources and proportionally more individuals.

The original rule invented by MacArthur was simple; if there are S species in an ecosystem, simultaneously break the line into S pieces by randomly picking S-1 points on the line and breaking the line at those points. The resulting lengths of the line segments can be described by a probability distribution, but it turns out to be far more uniform than are the typical SADs observed in nature. In other words, the distributions of line segments formed by this process are too tightly peaked around a modal length and they resemble neither a lognormal, nor a Fisher logseries, nor any other candidate for a realistic SAD. Pielou, Sugihara, and others modified the rule in various ways, including ways that result in distributions that resemble the lognormal or the more realistic truncated lognormal in which the mode is displaced toward low-log(abundance). Such an asymmetric distribution is called "left skewed" meaning that the mode is skewed leftward along the log(abundance) axis. The terminology is confusing because there are more species to the right of the mode than to the left in a left-skewed distribution.

In all the niche-based models that have been developed to explain SADs, including broken stick models, a fundamental problem concerns the relationship between the apportioned volume or segment of niche space assigned to a species and the abundance of the species. Perhaps species with more resources have larger individuals, not more individuals? Larger organisms tend to belong to species with fewer individuals. Could the original MacArthur version of the broken stick model be a reasonable distribution for resources, but not for abundances? Careful analyses of the consequences of assuming different causal relationships among resource use, body size, and abundance are given in Taper and Marquet (1996), and Gaston and Blackburn (2000).

Another way of introducing niches into ecosystem community models is with generalized Lotka–Volterra equations (Volterra 1926; Lotka 1956). These coupled, non-linear, time-differential equations provide a means of studying the dynamic consequences of choices of parameters describing regions of niche space occupied by species, as well as the time dependence and/or the steady state properties of either the number of individuals or the total biomass in each species. The parameters in the equations, in their most general form, characterize who eats whom and at what rate, the saturation kinetics of each trophic interaction, the competitive ability of each species paired against every other species, and the amount of carrying capacity available to each species. It is difficult to extract general principles about mass or abundance distributions from these equations because of the proliferation of parameters, although they have provided valuable insights into questions about diversity-stabilizing mechanisms, existence of population cycles, influence of stochastic environmental parameters, food web collapse under species deletions, and many other interesting ecological phenomena. Some of the more prominent examples of current niche-based, mechanistic models of ecosystem composition include the work of Hastings (1980), Tilman (1994), Hurtt and Pacala (1995), and Chesson (2000). A particularly good review is Berlow et al. (2004).

Finally, a recent and innovative synthesis of niche-based theory with power-law modeling has been developed by Ritchie (2010). He assumes that abiotic niches in an ecosystem have a self-similar structure in space, and uses a packing rule to allocate the regions of niche space to individuals and species. Thus the resources available to grow populations within species depends on the portion of niche space that species traits permit the species to occupy.

4.3.4 Island biogeographic theory

Profoundly influencing all subsequent theories of the distribution and abundance of species was the theory of island biogeography, originated by MacArthur and Wilson (1967). Their work provides a theory of independent-area SARs, applicable to either real islands or conceptual islands on mainlands; in what follows we call all such places islands. The essential idea is that an equilibrium number of species will be found on an island because colonization and extinction will at some level of species' richness come into balance.

The reasoning is quite robust and is based on a conceptual model in which a pool of potential propagules (that is, individuals that can reach the island) exists at some specified distance from the island itself. The species from which these propagules are found comprise the pool of potential immigrant species to the island. The theory posits that the rate at which new species arrive at, and establish on, an island depends on three factors: it will negatively correlate with the distance from the source pool to the island, it will correlate positively with the area of the island, and it will correlate negatively with the species' richness on the island because the chances that a random propagule from the mainland is a new species for the island decreases as the number of species on the island increases.

The theory also posits that extinction rates increase as the island gets more crowded. There are two reasons for this. First, if each species has some probability per unit time of randomly becoming extinct, the more species present, the more will go extinct in any time interval. Second, the more species and individuals present, the more likely that competitive interactions will drive some species extinct. The equilibrium number of species will be determined from the balance between extinction and colonization, and this balance will result in a number of species that will be some function of distance from source pool to island, the composition of the pool of colonizing species, and the area of the island.

Applying these ideas to an archipelago of many islands of differing size and distance, the actual shape of the equilibrium SAR will depend on the details of the dependence of the colonization and extinction rates on area. Area, alone, will not explain species' richness on the islands because distances also matter. Finally, the shape of the SAR will depend on the abundance distribution of the source pool of species because the colonization process is conceived of as a sampling of individuals from that pool (see Section 3.3.7).

An excellent review of evidence supporting this theory and its later incarnations is given in Gaston and Blackburn (2000). Many current theories owe much to the work of MacArthur and Wilson, less because of the details of the original theory and more because it identified, and folded into a predictive semi-quantitative model, a set of factors plausibly influencing biodiversity. Important to the model are time, area, dispersal distance, stochasticity, and source pool composition; niche parameters are relegated to seconday importance or ignored altogether in various offspring of the original theory. Thus this theory stands in contrast to the types of niche-based theories of the abundance and distribution of species described above. This distinction is useful today in categorizing models of macroecology. Theories like Hubbell's NTE, which assume dispersal limitation and neutral, stochastic demographics, have their intellectual antecedents in the work of MacArthur and Wilson.

4.4 Energy and mass distributions

Relatively little theoretical analysis has been directed at deriving energy or size distributions of individuals, as described by metrics such as $\Theta(\epsilon|n_0, N_0)$, $\Psi(\epsilon|N_0, S_0, E_0)$, and $\epsilon(n_0| N_0, S_0, E_0)$ in Table 3.1. The relationship between body mass and metabolic rate is the subject of metabolic scaling theory discussed in Sections 1.3.7 and 3.4.15, but because that theory does not predict a species–abundance distribution, it does not yield an actual probability distribution of body sizes across the individuals within a species or across the species in a community.

An evolutionary model of body mass diversification developed by Clauset and Colleagues (Clauset and Erwin, 2008; Clauset et al., 2009a; Clauset and Redner, 2009) yields both the steady state mass distribution across species (Eq. 3.20) and the historical rate of mass diversification as groups of related species diversify. The mass of a species refers here to the average mass of its individuals, just as it does in our metrics $\bar{m}(n)$ and $\mu(\bar{m})$ in Eqs 3.18 and 3.20. The basic modeling approach stems from physics models of diffusive branching processes and predictions are generated using convective reaction–diffusion models (Murray, 2002). Three premises are made. First, a species of mass m produces a descendent species with mass that is random multiple, λ, of m. The sign of the average of $\ln(\lambda)$ then determines whether the bias in mass evolution is toward larger or smaller individuals. From the fossil record, a tendency is seen for evolution to result in larger organisms, (a phenomenon called Cope's rule; Stanley 1973) and so the average of $\ln(\lambda)$ is taken to be positive. Second, in the Clauset model, extinction is a stochastic process for each species, independently, and at a rate that is weakly proportional to mass. Third, there is a minimum mass below which a species cannot exist. Evidence in support of each of these assumptions is cited in Clauset and Redner (2009).

From these assumptions, the metric $\mu(m)$ (Table 3.2) describing the number of species (in some assumed area) having mass m at time t is derived. When steady state is reached, a plot of log(number) of species versus log(mass) gives a curve that rises steeply and with negative curvature, reaching a broad maximum at about 100 times the minimum mass, and then, more slowly than it rose, it falls over about five decades of mass to give a number of largest species in a logarithmic mass interval that is below the number of smallest species in such an interval. The details of the shape will depend on choices of parameters, but the overall shape of the curve appears robust. Clauset and Redner (2009) show that it is also in reasonable agreement with data for terrestrial mammals.

4.5 Mass–abundance and energy–abundance relationships

The basic idea behind many models attempting to predict the relationship between the abundance of a species and the average mass (or metabolic rate) of its individuals, is that environmental resources, such as light or nutrients or water, in a fixed area, can only support some fixed amount of life. Whether it is the growth rate, or the standing crop of biomass, or total rate of metabolic energy use that is assumed not to exceed some fixed limit, depends on the model. But if this zero-sum assumption is accepted, then species' traits and/or stochastic factors allocate the available resources among them.

Models of that allocation process can be used to predict mass– and energy–abundance distributions across species. Recall that the broken stick model and its offspring cannot predict the SAD from first principles, but rather must assume a "breaking rule" that gives the desired answer. Similarly, models that allocate resources across abundance and body size with the goal of deriving the energy-equivalence rule (Section 3.4.10) or the related mass–abundance relationships (MARs) in one of its several manifestations listed in Table 3.3, have to assume some allocation principle and thus cannot derive the MAR from first principles. An important series of papers (Enquist et al., 1999; Enquist and Niklas, 2001) has extended metabolic scaling theory to ecosystem scale and used it to explore the connection between the energy-equivalence concept and mass–abundance relationships. A derivation of the energy-equivalence rule within the framework of the MST requires additional assumptions, unfortunately, which are arguably equivalent to assuming the answer. I will show in Chapter 7 that METE predicts the energy-equivalence rule under identified and very specific conditions on the state variables characterizing the ecosystem.

4.6 Food web models

There are many characteristics of complex networks, such as food webs, that could be investigated theoretically. These include not only our link metric, $\Lambda(l|S_0, L_0)$, discussed in Section 3.3.12, but also other measures such as the length of the longest chain connecting the base of the web to a top predator, the relative number of species at each trophic level, and the number of closed loops in the network. Early attempts to model the structure of food webs were carried out by various authors (Cohen et al., 1990; Yodzis, 1980, 1984; Pimm, 1982; Sugihara et al., 1989), usually basing their models on stochastic assembly rules or on filling of a pre-ascribed niche-space. A comprehensive review of food web structure and modeling efforts to understand it (Dunne, 2009) concluded that of all the early attempts, Cohen's cascade model was the most successful, although it too had some systematic failings. This stochastic model is based on a single state variable, the number of species in the web, with the number of links assumed to be proportional to the number of species. It distributes links between species by arraying all species uniformly in a linear hierarchy in which each species can only eat other species that are lower in the hierarchy. They do so randomly. While the model successfully described a composite average of patterns over many food webs, it could not adequately explain variations in webs with different values of the average number of links per species.

Many other models have been proposed (see review by Dunne, 2006), with arguably the most successful being the niche model of Williams and Martinez (2000). The niche model arrays species uniformly in a linear hierarchy and each species is also assigned a niche interval along that line, thereby loosening up the somewhat over-constrained cascade model. By assigning niche width stochastically for each species, the model improves on the cascade model in several ways, among which is that its output is more consistent with the observation that real food webs tend to obey the "constant connectance" rule (Martinez 1992), which states that the ratio of the number of links to the square of the number of species (i.e. fraction of all possible links that are realized) is the same in all webs.

Our link distribution metric $\Lambda(l|S_0, L_0)$ was not predicted by the niche model. Indeed, a form for that metric is effectively assumed at the outset; the amount of niche space that is consumed by a species is assumed in the niche model to be drawn from a beta distribution (Evans et al., 1993). I will return to this issue later, in the context of MaxEnt, but point out here that Williams (2010) addressed the question of link distribution using MaxEnt, without making any of the assumptions of the niche model. He predicted an exponential distribution (see discussion surrounding Figure 3.5a) for $\Lambda(l|S_0, L_0)$ and showed that it is in reasonable agreement with available data, although many exceptions exist. Because the exponential distribution is close to the beta distribution used in the niche model, MaxEnt effectively justifies that somewhat arbitrary choice of distribution made by Williams and Martinez.

4.7 A note on confidence intervals for testing model goodness

Statistical analysis of nested species–area data poses another interesting issue. When a linear regression is performed on a graph of $\log(S)$ vs $\log(A)$, one obtains both an R^2, providing a measure of the goodness of the regression model, and best-fitted values for the slope and intercept parameters in the model. In addition, your statistical package will generate for you confidence intervals on those values. You might desire that confidence interval so that you will know whether the slope, which is the power-law exponent, is significantly different from a value for the exponent that is predicted from some model. However, for nested census data, the confidence interval obtained for the exponent of the power-law SAR is not meaningful because the nested SAR is obtained using data that are not independent. For example, nestedness requires that the same data used to determine species' richness at 1 m^2 are being used to determine species' richness at 2 m^2 scale. This lack of independence of the data across spatial scales invalidates the information the statistical package gives you for the confidence interval on the slope of the $\log(S)$ vs $\log(A)$ plot.

Following Green et al. (2003), I describe in Box 4.3 a general method that can be used for such tests.

Box 4.3 Testing models of nested SARs

To accept or reject, say, a power-law or a random placement model for the SAR, you need to know how much the mean species.' richness at each scale can diverge from its expected value in landscapes for which each of these models holds. Hence, to test the fractal and random placement model species' richness predictions, you would simulate a large number (often 1000 in practice) of landscapes for each model. For example, with the RPM, you would run 1000 simulations of landscapes in which you assume a fixed probability $p = A/A_0$ for an individual being located in a cell of area A. With that rule, a set of species with abundances $\{n_0\}$ can be distributed on a landscape scaled down to area A. For each simulated landscape, the SAR can be constructed. For the fractal model (Appendix B) use of the probability parameters, $\{a\}$, or the parameter a, can be used to generate landscapes.

For each simulation, you can calculate and compare two statistics analogous to those commonly used in the analyses of spatial point patterns (Diggle, 1983; Plotkin, 2000):

$$k_{\text{simulation}} = \sum_i \left(\log\left(1 + \bar{S}_{\text{model}}(A_i)\right) - \log\left(1 + \bar{S}_{\text{simulation}}(A_i)\right) \right)^2, \qquad (4.22)$$

$$k_{\text{observed}} = \sum_i \left(\log\left(1 + \bar{S}_{\text{model}}(A_i)\right) - \log\left(1 + \bar{S}_{\text{observed}}(A_i)\right) \right)^2, \qquad (4.23)$$

where $\bar{S}_{\text{model}}(Ai)$, $\bar{S}_{\text{simulated}}(A_i)$, and $\bar{S}_{\text{observed}}(A_i)$ are the predicted, simulated, and observed values of mean species' richness at spatial scale A_i, respectively. The simulated value is the average over all simulations. The log-transform of the data is carried out because that generally increases the uniformity of the deviations around the mean of the residuals in the simulation. That is, it improves the homoscedasticity, a prerequisite for application of many statistical tools.

If k_{observed} is greater than a high percentage of the $k_{\text{simulation}}$ values, you may conclude that the model does not describe the observed data. Typically one uses one minus the percentage of simulations, where $k_{\text{observed}} > k_{\text{simulation}}$ as the p-value of our tests. Green et al. (2003) applied this method of significance testing to both the fractal model and the RPM, reaching the conclusion that neither model adequately described the serpentine plant census data.

4.8 Exercises

*Exercise 4.1

(a) Show that in the vicinity of its mean value, $\bar{n}$, the binomial distribution in Eq. 4.1, is well approximated by a Gaussian (or Normal) distribution: $\Pi(n|A,n_0,A_0) \propto \exp[-(n-\bar{n})^2 / (2\sigma^2)]$ where $\sigma = n_0 p(1-p)$.

(b) Show that when p in Eq. 4.1 is $\ll 1$, the binomial distribution in Eq. 4.1 is well approximated by the Poisson distribution: $\Pi(n|A,n_0,A_0) \approx [(pn_0)^n/n!] \exp(-pn_0)$, where $p = A/A_0$.

(c) Show that the mean value of the binomial distribution in Eq. 4.1 is given by:

$$\bar{n} \equiv \sum_{n=0}^{n_0} n\Pi(n|A, n_0, A_0) = pn_0 = n_0 \frac{A}{A_0}.$$

Explain in words why this had to be the case.

Exercise 4.2

Assuming the binomial distribution in Eq. 4.1, calculate the probability that, if $n_0 = 10^6$, that there are more than 5.1×10^5 individuals in $A = \frac{1}{2} A_0$.

*Exercise 4.3

Derive the dependence on A and D of the species-level commonality function $C(A, D|n_0, A_0)$ in the Coleman RPM.

*Exercise 4.4

Show that application of Eq. 4.3 results in the uniform distribution in Eq. 4.5 If you get stuck, look at Harte et al. (2005).

* Exercise 4.5

Derive the Laplace successional rule. Start with a simple case: assume there are just 3 balls in the urn initially, and all you know is that some number between 0 and 3 of them are red (R) and of course 3 minus the number of red ones is the number of blue (B) ones. There are four options for the initial mix: they can be RRR, RRB, RBB, or BBB. The order of the Rs and Bs in each initial mix is irrelevant. Assume the first ball selected is R: if you do not replace it, what is the probability that the next one you pick will be red, and what is the probability that all three balls are red?

(Hint: Consider the four intial possibilities for what is in the urn as listed above. With no additional information, each of these must have a probability of ¼: P(RRR) = P(RRB) = P(RBB) = P(BBB) = ¼. To proceed, you may want to use Bayes' Law, one form of which can be derived by noting two ways to write P(RRR): it is equal to P(RRR|R)P(R) and also to P(R|RRR)P(RRR). P(RRR|R) is the probability that if the first picked ball is red, then all three are red; this is what we want to determine. Similarly, P(R|RRR) is the probability that if the first three picked balls are red, then the first one is red (this is obviously equal to 1). P(R) is the unconditional probability (that is, without any prior knowledge) that a picked ball is red (clearly ½) and $P(RRR)$ is the unconditional probability that all three balls are red (¼ as explained above). Hence, equating the two ways to write $P(RRR)$, we have $P(RRR|R) = PP(R|RRR)P(RRR)/P(R) = (1)(1/4)/(1/2) = 1/2$. You can now show that $P(RRB|R) = 1/3$, and $P(RBB|R) = 1/6$. Then you can proceed to show that the probability that the second picked ball is red, given that the first is red, is $(1/2)(1) + (1/3)(1/2) + (1)(0) = 2/3$.)

Exercise 4.6

Show that the close agreement of the negative binomial distribution with $k = 1$ and the Laplace model are in close agreement for values of $n > 0$.

Exercise 4.7

Read Appendix B.1 and derive the relationship $\lambda_i(n_0, A_0) = 1 - \Pi(0|A_i, n_0, A_0)$ from Eq.B.1.1 and the definition of the a_j.

Exercise 4.8

Derive Eq. B.1.6 from Eq. B.1.5.

Exercise 4.9

Read Appendix B.4 and determine the dimensionless function $g(z)$ in Eq. B.4.11. (Hint: if the two patches share a common edge, so that $D^2 = A$ (recall that D is the distance between patch centers), then the number of species in common to the two patches, $N(A, A, A^{1/2})$, is exactly given by the difference between twice the number of species in a patch of area A and the number in the patch of area $2A$.)

*Exercise 4.10

Choose several species from Appendix A (the serpentine dataset) and perform on each the test described in Box 4.3 to determine which if any spatial distributions are consistent with the fractal model. In other words, determine whether $B(A)$ has power-law dependence on A.

Part III

The maximum entropy principle

In order to arrive at what you do not know
　You must go by a way which is the way of ignorance.

T. S. Eliot, "East Coker," from *The Four Quartets*

5

Entropy, information, and the concept of maximum entropy

Here we travel the path that leads from early nineteenth-century efforts to understand pistons pushing gases, to current ideas about inference and probability. Along the way I will explain the different meanings of thermodynamic and information entropies, and convince you that maximizing information entropy provides a means of implementing the agenda spelled out in Chapter 2. Specifically, it allows us to extend prior knowledge by finding the least biased distribution consistent with that knowledge.

Many of the grandest achievements in science have come about when disparate phenomena are found to be closely connected. For example, beginning at the end of the eighteenth century, the accumulating studies by Benjamin Thompson (Count Rumford), Sadi Carnot, Julius von Mayer, and others led to what we now take for granted but at the time was considered an astounding unification: the motion of objects and the warmth of objects were shown to be a single entity called energy. Thermodynamics, the science of energy, emerged. Equally profound was the nineteenth-century unification of electricity, magnetism, and light achieved by Michael Faraday and James Maxwell. Then, early in the twentieth century, the boundary between space and time disintegrated with Albert Einstein's theory of special relativity.

Less dramatic is another twentieth-century conceptual unification that sprang from Claude Shannon's creation of information theory (Shannon, 1948; Shannon and Weaver, 1949) and the work of Edwin Jaynes (1957a, 1957b, 1963). It resulted in the establishment of foundational connections between information theory and thermodynamics, between information and energy, and between physical entropy and a precisely defined metric of states of knowledge. The maximum entropy principle (MaxEnt) is an offspring of information theory. To explain MaxEnt, I need to explain some of the basics of information theory, and to do that I must begin with the concept of entropy in thermodynamics.

5.1 Thermodynamic entropy

Thermodynamics in general, and entropy in particular, can be approached and understood from many perspectives. One perspective reflects thermodynamics' origins in practical considerations of flames, pistons, and cylinders of gases.

Energy flow can be either reversible or irreversible. Picture a perfectly elastic rubber ball with no internal friction dropped in a vacuum from some height to a hard surface. The ball will drop, bounce, rise up, reach some maximum height and fall again. This process will continue forever because of the assumed condition of no friction. The ideal process is reversible, and in fact a movie of the ball bouncing up and down would look the same if it was run backward in time. Also, the process generates no heat exchange, again because there is no friction either within the ball, between ball and floor, and between ball and the vacuum through which it moves. While the velocity of the ball is changing, reaching zero at its apogee, its total energy, the sum of potential and kinetic, remains constant.

With this in mind, consider the formula defining a change in entropy (S), in terms of a change in heat content (Q) at temperature (T), that was originally introduced to science by Sadi Carnot in the nineteenth century:

$$\Delta S = \Delta Q/T. \tag{5.1}$$

In the case of our ball, there is no heat produced as it bounces up and down, and so the numerator is zero. Entropy is not changing. In general, for a system undergoing any reversible process, its entropy remains constant.

But now imagine the ball is made of moist clay. When it hits the ground it doesn't bounce. Instead, heat is produced and the energy in that heat will equal the kinetic energy the ball had when it struck the ground. Because heat is produced, $\Delta Q > 0$ and the the positive numerator in 5.1 implies entropy is produced. Moreover, the process is now not reversible. You can see this by supposing you had stored the heat produced when the lump struck the ground. Heating the clay and floor by that amount of heat would not cause it to rise up off the ground and resume its trajectory in reverse! And this is the essence of the original concept of entropy developed by people with a need to know how engines of various sorts work. You can convert motion to heat, conserving energy, but you cannot convert a unit of heat energy back into a unit of kinetic energy without some intervening process that will require additional energy. Irreversible processes produce entropy; reversible processes do not change it. In a closed system, entropy will not decrease.

Why "closed system"? If the system is open to outside influences, then there are ways to add energy to the system consisting of the clay splat and the floor and cause the clay to rise: for example, you could just walk up to the clay and, with a little work, lift it. The rule above about entropy not decreasing will apply only to closed systems. If enough external energy from outside the system is available, it can do almost anything, including create life and enhance its productivity and diversity over time, thereby decreasing the entropy of the more narrowly-defined system.

Another perspective from which we can understand entropy, a perspective closer to its information theory counterpart, is entropy as a measure of likelihood. To explore this we examine the likelihood of existence of different physical states. This will give us insight into why heat is so different from work or motion, even though both are forms of energy. Consider the molecules in the clay ball while it is falling.

Every one of them will have an identical velocity component in the downward direction due to the falling motion, and in addition, because of thermal energy, the molecular velocities will have a random component. In contrast, when the ball has hit the ground, only the random velocities will exist and they will be a little greater than before because of the generated heat upon impact.

We can think about this in terms of likelihoods if we first consider a simpler system: four molecules that can either be moving to the right or left with a unit of velocity. This is like a classic coin-toss problem applied to molecules moving left (heads, say) or right (tails). If the molecular motions are independent of each other, and the molecules are distinguishable, there is one way in which all four are moving to the left, one way in which all are moving to the right, four ways in which one can be moving to the left and three to the right, or vice versa, and six ways in which two can be moving to the left and two to the right. So a net velocity of zero is the most likely configuration. Similarly, if the molecules could be moving in any direction, not just left or right, the most likely configuration is one in which the velocity vectors add to zero. That state is less ordered than the one in which all molecules are moving to the left. Returning to our lump of clay, there are more ways for the molecules to all possess purely random velocity vectors than for there to be coherence in their velocities, with all having some common directional component, as well as a random component. In that sense, heat, which is the energy associated with the random movement of molecules, is more likely than coherent motion.

To further explore this, we have to relate the number of ways a system can be in a certain configuration to the likelihood of that configuration. Here we distinguish the microstate of the system from the macrostate. The macrostate of the clay object will be either a cool ball falling to the ground or, a little later, a slightly warmer ball lying on the ground. The microstate is the actual set of velocities, and positions, of all the molecules. Many microstates are associated with the ball on the ground, fewer with the falling ball because of the counting argument given in the paragraph above. Like that high-dimensional niche space that we discussed in Section 4.5, we can imagine now a very high-dimensional space, called phase space, in which each axis is an x, y, or z component of the velocity or the position of a molecule. In that space, the system at any moment can be described by a single point. Macrostates consist of a cloud of points because many microstates will be consistent with the same macrostate (for example, simply swap the velocities of two molecules moving in opposite directions). Some macrostates correspond to a bigger cloud of points than others, however, and in fact from the discussion above, the lump on the floor has more microstates associated with it than does the falling ball. The likelihood of a macrostate is proportional to the number of microstates associated with it, and the entropy of a macrostate, S, is related to the number, W, of microstates it can be in by Boltzmann's famous formula engraved on his tombstone:

$$S = k \log W. \tag{5.2}$$

The multiplicative constant k is called Boltzmann's constant.

Without going in to the mathematical details and the interesting historical nuances, which can be found in good texts on statistical mechanics or thermodynamics (e.g. Grandy, 1987, 2008), Boltzmann and Gibbs showed the following. Consider a system (often called a thermodynamic system) consisting of N gas molecules, each of which can have energy ϵ_i, i = 1, ..., M. Let $p_i = n_i/N$ be the fraction of molecules with energy ϵ_i. Then from Eq. 5.2, the entropy of the system can be written:

$$S = -k \sum_{i=1}^{M} p_i \log(p_i). \qquad (5.3)$$

Maximizing this entropy measure, subject to the constraints that $\Sigma_i n_i = N$ and $\Sigma_i n_i \epsilon_i = E$ gives the Boltzmann distribution of molecular energy levels in an ideal gas. The rest of classical thermodynamics follows.

The concept underlying all this is that the probability a system will be found in a particular macrostate increases with the number of microstates associated with it. By analogy, think of wandering truly aimlessly (like a molecule) through a mall that contains 10 toy stores and only 1 candy store... you will spend more time in one of the former than the latter. Similarly, if you are in a macrostate today with M microstates, the chances are that tomorrow your macrostate will contain at least M microstates; the chances it will contain significantly fewer are vanishingly small. Unusual systems transition to become more usual systems, not vice versa. An isolated bar of metal that is hot at one end and cool at the other is, on its own, going to become uniformly luke warm, rather than extra hot at the hot end and extra cold at the cool end. A moving ball of clay can generate heat more readily than a hot ball of clay can spring into motion.

To summarize, more likely macrostates are states of higher entropy because they are compatible with more microstates. Systems undergoing change are more likely to transition to more likely, not less likely, macrostates states and thus are more likely to increase entropy. The second law of thermodynamics is thus a probabilistic statement.

As an aside, some popularizations of the entropy concept refer to high-entropy states as always being "disordered" and low-entropy states as being more highly ordered. This is often the case, but it is strictly speaking not correct, and in any case the concept of order is difficult to define objectively... unless, in circular fashion, one defines it in terms of entropy. Now it is true that if the barrier between containers of two different gases is opened, the gases in each container expand into the other container, the molecules mix, and the system becomes more disordered by any reasonable definition of order. But the change in entropy of the system is a result of the expansion of each of the gases into the doubled volume available to the molecules, not to the mixing (disordering) itself (Ben-Naim, 2006).

To see the subtleties of attempting to associate entropy with disorder, consider the case of two barrels on the back of a truck on a bumpy road. One barrel starts out containing a tall stack of plates, with the widest on the bottom and with plate diameter decreasing with height; the other contains a large number of pencils randomly thrown in to the barrel. After a short drive, the first barrel contains a more disordered set of plates, while the second contains somewhat more aligned pencils. Disordering of the plates is expected. Alignment of the pencils occurs because there are more "microstates" possible per unit volume as the pencils align and become more densely packed. For further discussion of this, see Grandy (2008).

5.2 Information theory and information entropy

So now you know what entropy is from two perpectives: Eq. 5.1 and in terms of likelihoods. We are ready to turn to information theory. It is important to understand at the outset that the term "information" is used in information theory in a rather narrow way. It does not have any direct connection to meaning, insight, or understanding. Rather, more information means a reduction of ambiguity, an increase is certainty. Your information, say a written message, could be totally counterfactual, but it would still be information if it can reduce ambiguity. I write "can reduce ambiguity" rather than "does reduce ambiguity," because the message might be written in a language that you cannot read!

When something unusual happens, such as when a person, whom you thought was far away, walks in your door, you gain a good deal of information. On the other hand, when something that is expected happens, such as someone you live with walks in your door, you have not learned much. Information has something to do with unusualness. The word game "hangman" illustrates this well; if I am trying to figure out what word you are thinking of, and if all I know is where the most common letters appear in the word, I know relatively little. For example, "_u _ _le" could be many different words, whereas if I know "p_ zz_ _" I would probably guess "puzzle" or "pizzas."

A real message with rarely used letters (e.g. z, q, x, j in English) is somewhat like a thermodynamic system that occupies a small cloud of microstates in phase space: if you know the macrostate of that thermodynamic system, you are better able to guess the microstate because fewer microstates are associated with it. High information content is like low entropy.

Now let's get quantitative. Consider an imaginary language written in an alphabet with 6 letters appearing with equal frequency in real messages written in that language. Moreover, assume that there are no correlations among the letters as they appear in real words. The letters are like the numbers on a fair die. If all I am told is that in some lengthy message the first letter of the alphabet appears in the 11th position, I have reduced by a little bit my confusion about the content of the message. The a priori probability that the first letter of the alphabet appeared in that position is 1/6. If I am told the above, plus the fact that in the 3rd position there appears the second symbol of the alphabet, then I have reduced my confusion by twice as much, and the *a priori* probability of both these pieces of information is $1/6 \times 1/6 = 1/36$. (Note that if there were correlations in letter occurrence, such as q always followed by u, then I would not necessarily learn twice as much.) So the product of the probabilities gives us a doubling of the amount by which my confusion is reduced. If I define a function $I(p)$ expressing the information gained by knowing something that had probability p of being "true" (recall, we are not dealing with actual augmentation of wisdom but rather with the conveyance of data), then we have:

$$I(p) + I(p) = I(p \cdot p). \tag{5.4}$$

This equation has a solution:

$$I(p) = K \log(p), \tag{5.5}$$

which is unique up to an overall constant K (which could be positive or negative) or, equivalently, up to an arbitrary base for logarithmic function.

More generally, let the language have L distinct letters and assume that from examination of volumes of text in that language the letters appear with frequencies p_i where $i = 1, \ldots, L$. In a lengthy message N-letters long, the ith letter of the alphabet can be expected to show up $n_i = p_i N$ times. Independently, Claude Shannon and Norbert Weiner showed that the language's capacity to inform, per symbol in a message, could be expressed as:

$$I_p = K \sum_{i=1}^{L} p_i \log(p_i). \tag{5.6}$$

I use the notation I_p rather than $I(p)$ to reduce the clutter in equations that will follow.

Referring back to Eq. 5.5, Eq. 5.6 makes sense: the information content of a set of letters is the expectation value of the information content of each letter and the sum of p times $\log(p)$ gives that expectation value.

Because we would like information to be a non-negative quantity, and because log $(p) \leq 0$ if $0 \leq p \leq 1$, we take K to be negative. It is also standard practice to measure information in "bits" and to take the information content of a single binary choice to be 1 bit. Hence, taking $p = 1/2$, corresponding to a binary choice such as a coin toss, we have $K \log(1/2) = 1$ bit. Therefore, in units of bits, it is customary to take $K = 1/\log(1/2) = -1/0.693\ldots \sim -1.44$.

The connection between information theory and thermodynamics starts to come into focus if Eq. 5.6 is compared to the Boltzmann/Gibbs result, Eq. 5.3. The mathematics of the likelihood interpretation of thermodynamic entropy bears a striking similarity to the expression for information content. Because the Boltzman constant k is positive, thermodynamic entropy is non-negative. A non-negative measure of information entropy is then plausibly given by:

$$\text{Information Entropy} \propto I_p. \tag{5.7}$$

At this point you might be understandably confused. We have asserted that information I and information entropy S_I are both non-negative. But from our discussion of thermodynamic entropy it is clear that high-entropy states are states in which we know very little, states in which there are many microstates and thus our knowledge of which one the system is in is quite small. In a high-entropy system we have very little information about microstates. But a careful definition of what is meant by information clears up the issue. Equation 5.6 describes the information that can be gained by a measurement (e.g. knowing another letter in the game hangman). In a low-entropy system, such as a room in which all the molecules are in one side of the room, measurement of the location of a molecule adds little to our prior knowledge. So information entropy is a measure of our confusion about the state of the system. The amount of information we have about a system prior to a measurement is sometimes referred to as neg-entropy; high-entropy states are states of confusion, which is another way to say that they are states in which the information we can gain about the system is large.

Importantly, the concept of information entropy can be applied to arbitrary probability distributions, not just to those describing the frequency of letters in a language or faces of a die. Consider, for example, the probability that a species selected at random from an ecosystem containing a total of N_0 individuals and S_0 species, has n individuals. This is what we referred to as the SAD in Table 3.1, $\Phi(n|S_0, N_0)$. From Eq. 5.6, an information theoretic measure of the information entropy of the distribution is:

$$I_\Phi = -\sum_{n=1}^{N_0} \Phi(n) \log(\Phi(n)). \qquad (5.8)$$

Note that because the constant out in front of the summation in Eq. 5.6 is actually arbitrary, we have simply set it equal to 1.

You might wonder why the summation in Eq. 5.8 is over all possible individuals rather than over all the species: the reason is that by summing over n (which is the domain of the probability distribution Φ) the product of (the probability of a species having n individuals) × (the information content in the choice of a species with n individuals), we are in effect summing over the species. This is similar to what is done in classical statistical mechanics, where in the derivation of the Boltzmann energy distribution over molecules, $C \cdot \exp(-E/k \cdot T)$, an integral is carried out over all possible energy states of a molecule; species are to molecules as abundances are to energy values.

5.3 MaxEnt

From Eq. 5.8, we can extract qualitative insight into the meaning of the "information entropy of a probability distribution" by examining some limiting cases. Suppose, first, that the probability distribution Φ is uniform... every value of abundance is equally likely. Then $\Phi = 1/N_0$ and because there are N_0 terms in the sum, $I_\Phi =$

$(N_0/N_0) \log(N_0) = \log(N_0) > 0$. Now take another case, in which Φ is zero unless $n = \tilde{n}$, in which case $\Phi = 1$. In this case, all species have the same abundance, which is also the average abundance per species. Recalling that $x \log(x) \to 0$ as $x \to 0$, and that $\log(1) = 0$, we see that $I_\Phi = 0$ when all abundances are equal. So the information entropy is greater for the case of a uniform probability distribution than for a sharply peaked distribution. In fact, it can be shown that information entropy is maximum when the distribution is uniform (Jaynes, 1982). This is consistent with our interpretation of information entropy in Eq. 5.7: it is a measure of our ignorance about the state of the system. We know least about a variable when the probability distribution for that variable is most uniform.

As discussed in Chapter 2, if we want to infer the shape of a probability distribution in such a way as to avoid bias, we should not assume anything about it that we have no basis for assuming. That might suggest that we should always assume that unknown probability distributions are uniform: $P(x) =$ constant. For example, if we have a 6-sided die, and have no reason to suspect the die is unfair (i.e. there is no obvious unevenness to its mass distribution), and we have no reason to believe that it is being tossed in an unfair way (e.g. always tossed from 5 cm above a table top with face 6 initially pointing upward), we would have no reason to assume anything other than that each face has a probability of 1/6 of coming up on the next toss. Suppose we do obtain some data about the die ... perhaps someone tossed it 10,000 times and determined that the average value is indeed $(1 + 2 + 3 + 4 + 5 + 6)/6 = 3.5$. This is consistent with the notion that the distribution is uniform, but it is also consistent with the possibility that $p(1) = p(2) = p(5) = p(6) = 0$ and $p(3) = p(4) = 1/2$. Which should we assume? Logically, it makes no sense to assume that only 3s and 4s will appear when we have no reason to suspect that is the case.

On the other hand, if we were told that the average value of 10,000 tosses was not 3.5 but rather 4, we would no longer be able to maintain our fallback position that the distribution was uniform. Now our intuition might guide us to ask the question: what is the most uniform distribution that the distribution could have that is compatible with the average value being 4, not 3.5?

In short, we want to infer the shape of the distribution by either of the following qualitatively equivalent operations:

(a) Maximizing our residual confusion after taking account of what we know.
(b) Finding the most uniform distribution subject to the constraints of our prior knowledge.

By either of these operations we will arrive at the least biased inference of our distribution. Statement (b). leads to a specific mathematical procedure: maximizing information entropy subject to the constraints imposed by prior knowledge and that is what is meant by MaxEnt. The actual mathematical procedure is called the "Method of Lagrange Multipliers" and it is explained in Box 5.1. It is a procedure that allows us to find the unique functional form of probability distributions, such as $\Phi(n)$ in Eq. 5.8 that maximizes I_Φ.

Box 5.1 Maximizing information entropy by the method of Lagrange multipliers

We seek the least biased estimate of the functional form of a probability distribution $p(n)$ that is subject to a set of K constraints that arise from prior knowledge and that can be expressed in the form of K equations:

$$\sum_n f_k(n)p(n) = \langle f_k \rangle. \tag{5.9}$$

Here, the averages (over the distribution p) of the functions f_k are denoted $\langle f_k \rangle$; they are the constraints, whose values we know from prior measurement or other sources of knowledge. One additional constraint is the normalization condition:

$$\sum_n p(n) = 1. \tag{5.10}$$

The independent variable n is summed over all its possible values, and the index k runs from 1 to K. It was shown by Jaynes (1982) that the solution to our problem is the function $p(n)$ that maximizes the "information entropy:"

$$I_p = -\sum_n p(n) \log(p(n)), \tag{5.11}$$

subject to those constraints. Maximization is carried out using the method of Lagrange multipliers. The maximization procedure yields:

$$p(n) = \frac{e^{-\sum_{k=1}^{K} \lambda_k f_k(n)}}{Z(\lambda_1, \lambda_2, ..., \lambda_K)}, \tag{5.12}$$

where the K quantities λ_k are called Lagrange multipliers and Z, the partition function, is given by:

$$Z(\lambda_1, \lambda_2, ..., \lambda_K) = \sum_n e^{-\sum_{k=1}^{K} \lambda_k f_k(n)}. \tag{5.13}$$

The λ_k are formally given by the solutions to:

$$\frac{\partial \log(Z)}{\partial \lambda_k} = -\langle f_k \rangle, \tag{5.14}$$

although in practice the Lagrange multipliers are usually most simply determined by starting with Eqs 5.9 and 5.13, with Eq. 5.12 substituted in for $p(n)$, and either algebraically or numerically solving the resulting simultaneous algebraic equations.

The form of Eq. 5.13 ensures that the form of $p(n)$ in Eq. 5.12 is properly normalized, so that Eq. 5.10 holds.

For a rigorous explanation of why Eq. 5.12 is the form of $p(n)$, that maximizes information entropy, consult an advanced calculus or applied mathematics text. Here is a hand-waving way to understand the solution. Consider just the two constraints, the normalization constraint in Eq. 5.10 and:

Box 5.1 (Cont.)

$$\sum_n f(n)p(n) = \langle f \rangle. \tag{5.15}$$

Maximizing I_p in Eq. 5.11 is equivalent to maximizing the expression:

$$W = -\sum_n p(n)\log(p(n)) - \lambda_0\left(\sum_n p(n) - 1\right) - \lambda_1\left(\sum_n f(n)p(n) - \langle f \rangle\right), \tag{5.16}$$

If the normalization condition and the constraint Eq. 5.15 both hold. The point is that if the constraints hold, then the terms multiplying λ_0 and λ_1 both vanish and so for any value of the λ_i we seek the maximum of W. To maximize a functional (that is, a function of a function), W, of $p(n)$ by varying the form of $p(n)$, we have to set the derivative of W with respect to p equal to zero. For each value of n, the derivative is:

$$dW/dp = -\log(p(n)) - 1 - \lambda_0 - \lambda_1 f(n), \tag{5.17}$$

and setting this to zero gives:

$$p(n) = k \cdot e^{-\lambda_1 f(n)}, \tag{5.18}$$

where:

$$k = e^{-(\lambda_0+1)} \equiv \frac{1}{Z} = \frac{1}{\sum_n e^{-\lambda_1 f(n)}}, \tag{5.19}$$

is a normalization factor. Equation 5.14 now follows from the actual constraint. In particular:

$$\frac{\partial \log(Z)}{\partial \lambda_1} = \frac{1}{Z}\frac{\partial Z}{\partial \lambda_1} = -\frac{1}{Z}\sum_n f(n)e^{-\lambda_1 f(n)} = -\langle f \rangle. \tag{5.20}$$

While our derivation was for the case of a single constraint along with a normalization condition, Eqs 5.12–5.14 hold for any number of constraints. It is also straightforward to generalize the derivations above to joint probability distributions $p(n, m, \ldots)$.

In the case of a continuous distribution, $p(x)$, sums over n are replaced by integrals over the continuous variable, x. In that case, a more general procedure is to maximize the expression:

$$I_p = -\int dx\, p(x) \log[p(x)/q(x)],$$

where $q(x)$ is a so-called reference distribution; it is a prior distribution relative to which you seek the posterior distribution $p(x)$. Situations in which it is justified arise sometimes for the case of continuous distributions in which the requirement can be imposed that the form of the probability distribution should be independent of the units used to express numerical values of quantities. Units-independence forces an additional constraint that can be dealt with using the reference distribution. Jaynes (1968, 2003) discusses this in depth. In our applications of MaxEnt to macroecology in Chapter 7, we will not have to deal with reference distributions because our distributions are either discrete or in the case of an energy distribution, the problem of units-independence will not arise.

Box 5.2 discusses some alternative measures of information entropy, but the one above, based on Shannon information theory, is the one we will largely focus on in this book.

Box 5.2 Alternative measures of information entropy

What are the general properties that a measure of information entropy should possess? If we can state such properties mathematically, in the form of axioms, then we can decide whether Eq. 5.7 satisfies them and if any other measures of information do so as well. We follow closely here the excellent treatment of this topic in Beck (2009).

Consider a discrete probability distribution $p(n)$, properly normalized to 1, and an information entropy, I_p, that is a functional of p.

Axiom 1. I_p should depend only on p, and not on other factors, such as the temperature or time of day. The function $p(n)$ will depend, conditionally on other quantities such as state variables, but those quantities do not influence the form of the dependence of I_p on p.

Axiom 2. I_p is maximum when $p =$ constant. In other words, the uniform probability distribution maximizes I_p. Recall, however, that a uniform p may not be possible because of the constraints on p.

Axiom 3. If the sample space of outcomes is enlarged, but the probability of events in that additional sample space is zero, I_p is unaffected. For example, if n is originally taken to run from $n = 1, 2, \ldots, N$, and I_p is computed, its value won't be influenced by extending the sample space to run from $1, \ldots, N, \ldots, N+M$, if $p(n) = 0$ for $n = N+1, N+2, \ldots, M$.

Axiom 4. This axiom states that in a situation in which information is sequentially obtained, the order in which information is acquired should not affect the information content at the end.

To understand Axiom 4, consider Eq. 5.4 generalized to non-independent outcomes. In that more general case, consider two systems, one described by $p(n)$ and the other by $q(m|n)$. Let the probability of finding a particular outcome, n' in the first system and m' in the second system be $r(n', m')$. Then the axiom states:

$$I_r = I_p + \Sigma_n p(n) I_{(q(m|n))}. \tag{5.21}$$

These four axioms were first introduced by Khinchin in 1957. They are easily shown to be satisfied by Eq. 5.4. In fact, it has been proven that the $-\Sigma p \log(p)$ form of information entropy is the unique form that satisfies the four axioms.

A wider class of information measures is possible if we relax one or more of the axioms. If we keep Eq. 5.4 but do not require Eq. 5.21 to hold, then a wider class of entropy functionals, called Renyi entropies, is possible. They take the form:

$$I_{\text{Renyi}} = \frac{1}{q-1} \log(\sum_n [p(n)]^q). \tag{5.22}$$

As q, an adjustable parameter that can be any real number, approaches 1, this entropy approaches the Shannon/Jaynes information entropy. Renyi entropy has been used in the study of mathematical structures called multifractal sets but appears to be of little interest in statistical mechanics.

Relaxing the necessity of Axiom 4 and not requiring Eq. 5.4, we arrive at the Tjallis entropy measure:

$$I_{\text{Tjallis}} = \frac{1}{q-1}(1 - \sum_n [p(n)]^q). \tag{5.23}$$

Box 5.2 (*Cont.*)

In contrast to the Renyi entropy, this measure has been used to attempt to generalize statistical mechanics. It, too, reduces to the Shannon/Jaynes measure as q approaches 1.

There are many others in the literature, some of which are discussed in Beck (2009). Of them all, the one that could potentially be of interest in macroecology is the Tjallis entropy and in Chapter 8, I return briefly to it.

The MaxEnt machinery in Box 5.1 implements the logical framework for inference laid out in Chapter 2. Equations 5.11–5.13 (Box 5.1) can be thought of as a recipe for drawing least biased conclusions in the face of uncertainty. By finding the distribution with the highest information entropy that satisfies known constraints, we achieve the least biased inference of the form of the distribution. In a colloquial sense, the lesson of this chapter is that we can be most certain of our inferences if we accept most fully our ignorance, which means not pretending to have knowledge that we lack.

If we refer back to our discussion of Laplace and the principle of indifference in Section 4.1.2, we see that the intellectual antecedents of the work of Jaynes go back well before Shannon and Gibbs to Laplace. That master of physics and mathematics, arguably the greatest since Newton, understood and applied the very same ideas that constitute the MaxEnt approach to inference.

Laplace applied Newtonian mechanics to resolve ambiguities in the orbits of Saturn and Jupiter and to predict the mass of Saturn. Conceptual problems originating with sparse data had prevented some of the greatest mathematicians of the eighteenth century from solving this problem. Using the limited orbital data as constraints, he was able to apply the principle of indifference to correctly infer a probability distribution for Saturn's mass and for alternative orbital solutions to Newton's equations.

Interestingly, Laplace was a "Bayesian," as was Jaynes. That means that they both rejected a narrow "frequentist" notion of the meaning of probability in favor of interpreting probabilities more broadly as a description of a state of knowledge. The description of this view of probability is well described in Jaynes (2003). In the frequentist view, the meaning of probability is tied to knowledge of the frequency of occurrence of events. The absence of frequency data is informative to a Bayesian, but not to a frequentist. Bayes' law, which we deploy in this book (e.g. Box 3.1, Exercise 4.5, Sections 9.4 and 11.1.2), is part of the machinery of implementing an approach to probability theory that liberates it from knowledge of frequency.

For an excellent review of these issues see Sivia (1993).

We have traversed a broad terrain on which remarkable and diverse events in the history of science have played out. We have gone from thinking about ambiguities in calculations of planetary orbits and the principle of indifference, to seemingly mundane but actually profound questions about steam engines; from the likelihood

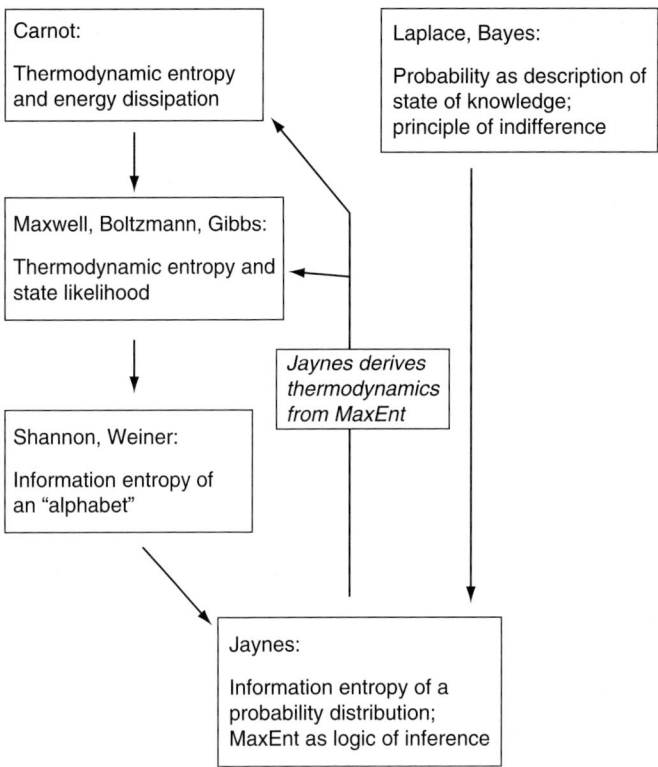

Figure 5.1 MaxEnt in historical context.

of states of molecular motion to the information conveyed by a message; and from unification of the diverse forms of energy to development of a logic of inference. It may be helpful to see the aerial view of this terrain in Figure 5.1.

And the end of all our exploring
Will be to arrive where we started
And know the place for the first time.

T.S. Eliot, "East Gidding," from *The Four Quartets*

6

MaxEnt at work

Here I look at practical aspects of MaxEnt in application. What does it mean to say it "works" or it "fails"? What do we conclude when it fails? What are common examples of its predictions? How has it actually been used?

From the way I derived MaxEnt in Chapter 5, you may be left with the impression that, since it is a logic-based method of obtaining least-biased knowledge, that it cannot fail; that it must always yield predictions that resemble nature. But in fact MaxEnt need not always "work." To understand what it means to say it "works," it helps to begin with a discussion of the ways MaxEnt can fail to yield reliable predictions.

6.1 What if MaxEnt doesn't work?

Suppose an application of the MaxEnt machinery to a real-world problem fails, in the sense that it does not make an adequate (to be defined by you) prediction. There are four ways in which MaxEnt, used in conjunction with Shannon/Weiner information entropy and the Jaynes formulation of the procedure, as summarized in Box 5.1, can fail to work.

First, there is the possibility that you made a mathematical error in working out the solutions to the equations in Box 5.1. We won't pursue that.

Second, one or more of the four Khinchin axioms (Box 5.2) may not be valid for your problem, and you should be using a different measure of entropy. This may be more of a problem than practitioners of MaxEntology let on. The fourth axiom is fairly strong, for it asserts that the order in which we gain information does not matter. It is not difficult (Exercise 6.1) to think of counterexamples.

Third, it is possible that your prior knowledge is not factually correct. If one of your constraints is the mean value over the sought distribution, and you miss-measured it or miscalculated it from your prior knowledge of state variables, or the values of your state variables were misestimated (see Chapter 7), your predictions will surely suffer as a consequence.

Fourth, you may have ignored some additional information in the form of constraints on your sought-after distribution. This is a pervasive and a subtle issue. Let's assume you possess certain prior knowledge in the form of constraints, and you choose to, or unwittingly, ignore some knowledge of additional constraints. Then, if the first three reasons for failure listed above do not apply, you will indeed obtain the best possible inference as to the shape of the distribution . . . given your willful or careless neglect. In other words, MaxEnt will work only as well as you work with it. Your least-biased estimate of something may be completely at odds with nature.

Ultimately there can be no solely logic-based path to knowledge of nature. When a theory that incorporates MaxEnt does yield reliable predictions, the logic of inference is certainly abetting that success, but the way in which you conceptualize both the type of sought-after distributions and the constraint information will play a huge role. MaxEnt does not eliminate from scientific pursuit the role of good intuition, common sense, careful data acquisition, and accurate calculations.

6.2 Some examples of constraints and distributions

Table 6.1 shows examples of the probability distributions that MaxEnt predicts from specific forms of the constraint function $f(n)$. Where two fs are listed, such as n and n^2, the implication is that two constraints hold simultaneously. Frank (2009) provides a useful overview of MaxEnt-derived distributions.

Let's work a couple of examples to ensure that you understand how to implement the mathematics in Box 5.1. First, take the case of a fair three-sided die. We wish to infer the values of $P(1), P(2), P(3)$ from prior knowledge that the mean value of n is $(1+2+3)/3 = 2$. Let's do it by brute force: Eq. 5.10 informs us that:

$$P(1) + P(2) + P(3) = 1. \tag{6.1}$$

The statement that the mean value is 2 can be expressed as:

$$1P(1) + 2P(2) + 3P(3) = 2. \tag{6.2}$$

Label the Lagrange multiplier λ. Then by Eq. 5.12:

$$P(n) = \frac{1}{Z}e^{-\lambda n}, \tag{6.3}$$

Table 6.1 Examples of MaxEnt solutions.

Constraint function f(n)	Form of $P(n)$ from MaxEnt
n	$e^{-\lambda n}$
special case: n = all integers between n_{min} and n_{max}; mean = $(n_{max}+ n_{min})/2$	$1/(n_{max} - n_{min} +1)$
(e.g. 6-sided fair die)	(e.g. $P(n) = 1/6$)
n, n^2	Gaussian (normal) distribution
$\log(n), \log^2(n)$	Lognormal distribution
$\log(n)$	$n^{-\lambda}$ (i.e. power law)

where:

$$Z = \sum_{n=1}^{3} e^{-\lambda n}. \qquad (6.4)$$

This problem is simple enough that we can dispense with Eq. 5.14. In particular, letting $x \equiv e^{-\lambda}$ for convenience, Eqs 6.1 and 6.2 can be combined to give:

$$x + 2x^2 + 3x^3 = 2(x + x^2 + x^3). \qquad (6.5)$$

This can be re-written as:

$$x^3 = x, \qquad (6.6)$$

which has three solutions: x = 0, 1, −1. These correspond to $\lambda = \infty, 0, i\pi$. Only the second of these corresponds to a real-valued, well-defined probability function, when substituted into Eq. 6.3. Hence:

$$Z = 1 + 1 + 1 = 3 \qquad (6.7)$$

and

$$P(n) = \frac{1}{3} e^{-0n} = \frac{1}{3}. \qquad (6.8)$$

A similar result obtains for an N-sided fair die: $P(n)$ is given by $1/N$.

Now suppose that in the case of the 3-sided die, you had prior knowledge that the average value of n was 1.5, not 2. Now Eq. 6.4 becomes:

$$x + 2x^2 + 3x^3 = 1.5(x + x^2 + x^3) \qquad (6.9)$$

or,

$$1.5x^3 + 0.5x^2 = 0.5x. \qquad (6.10)$$

Dividing out the x = 0 solution, which again is impossible, gives:

$$1.5x^2 + 0.5x = 0.5. \qquad (6.11)$$

This has two solutions: $x = (-3 \pm 13)/6$. Only the solution with the plus sign (~ 0.101) gives a real-valued λ (~ 2.29) and thus $Z \sim 0.112$; $P(1) = 0.9$; $P(2) \sim 0.090833$; $P(3) \sim 0.009167$.

Had the constraint value for the mean of n been > 2, then the solution would have $\lambda < 0$, and $P(n)$ would be an exponentially increasing function of n (Exercise 6.2). In general, for the case in which there is a single constraint and it is the mean value that is known, $P(n)$ will either be uniform, if the mean $= (n_{max} + n_{min})/2$, or exponentially decreasing, if mean $< (n_{max} + n_{min})/2$, or exponentially increasing, if mean $> (n_{max} + n_{min})/2$.

You can readily see how the various other entries in Table 6.1 arise. Take the case of a constraint on $\log(n)$. Because $e^{-\lambda \log(n)} = n^{-\lambda}$, a power law form for $P(n)$ results immediately.

6.3 Uses of MaxEnt

In Chapter 5, I mentioned that the first application of the concept of MaxEnt can arguably be traced back to Laplace. It was only with the work of Jaynes, however, that MaxEnt became more formally described (as in Box 5.1) and more widely accessible, understood, and used. Here I summarize more recent applications.

6.3.1 Image resolution

A large class of uses looks slightly different at first glance from thinking about loaded dice or planetary orbits, but boils down to the same logic. These are the applications that can generally be called image-resolution problems. Faced with a fuzzy image in medicine or forensics, MaxEnt has been used to infer the most likely detailed image, subject to the constraints imposed by the available image. An excellent summary of how and why MaxEnt is used toward that end is found in the collections of contributed chapters in Buck and Macauley (1991), with the chapter by Daniell (1991) particularly useful.

6.3.2 Climate envelopes

Related to image reconstruction is the problem of inferring what are called "climate envelopes" for species. Under global warming, we would like to know how the climatically-suitable habitat for each species will shift, perhaps poleward and/or uphill. To accomplish that we need to know how the likelihood of presence of a species at any particular location depends on the climate at that location. How is that done? Suppose you overlay a relatively fine grid over a large biome, and have some way of knowing the climate conditions in each grid cell. Suppose, further, that from incomplete (not every cell is necessarily sampled) censusing you have data informing you as to whether a particular species is present in the censused cells. If it is a very intensive census, the data set may also inform you as whether the species is absent in those cells, although sometimes census-takers have much more confidence in their presence data than in their absence data. The task of constructing a climate envelope is the task of finding a probability distribution that estimates the likelihood of presence.

The application of MaxEnt to that task was first proposed and developed by Phillips et al. (2004), with subsequent extensions of the method (e.g. to presence-only data) by Elith et al. (2006), Phillips et al. (2006), and Phillips and Dudik (2008). MaxEnt is one of many methods, however, that has been used to accomplish that. Useful inter-comparisons of all the commonly used approaches are given by Phillips and Dudik (2008), and Elith and Leathwick (2009). The basic idea of the MaxEnt approach is probably anticipated by the reader: MaxEnt finds the smoothest possible probability distribution for a species' occurrence over geographical sites that satisfy constraints derived from the combination of climate data and data on observed presences, or observed presences and absences if absence data inspire confidence. Numerical algorithms are used to carry out the procedure (see, for example, Phillips et al., 2006). I will return to this topic in Chapter 9, where I illustrate with a simple example how MaxEnt can be used to infer climate envelopes.

6.3.3 Economics

Two economists, Amos Golan and George Judge, have pioneered the application of MaxEnt to a variety of topics in the field of economics (Golan et al., 1996; Judge et al., 2004). Typical of the kind of problem that they and their colleagues have addressed is the so-called "ecological inference" problem that arises in economics and also in political science and sociology (see, for example, King, 1997). Here is a generic example of this problem. In an election, it is known that three political parties (A, B, C) divided the total vote 15%: 30%: 55%. Moreover, it is known that the total number of voters in each quartile of income were divided (45%, 25%, 20%, 10%). The question is: what can be inferred about the fraction of voters in each quartile who voted for each party? In other words, the problem consists of filling in the elements of a matrix when the row and column sums are specified (Table 6.2).

A similar type of problem arises in economics with input–output tables of an economy. Input–output tables or matrices are designed to portray the flows of economic value between sectors of the economy. For example, it takes steel to make a coal-fired power electric-generating plant and it takes electricity to make steel. So, if power generation and steel making were sector labels on both a row and a column, then the $m_{steel,electricity}$ entry could be the value of the steel used by the

Table 6.2 An example of a voter preference table. Row sums are percent of voters in income quartiles; column sums are precent of voters supporting each of three political parties.

Income quartile/party	A	B	C	Sum
1	m_{1A}	m_{1B}	m_{1C}	45
2	m_{2A}	m_{2B}	m_{2C}	25
3	m_{3A}	m_{3B}	m_{3C}	20
4	m_{4A}	m_{4B}	m_{4C}	10
sum	15	30	55	

power sector, and the $m_{electricity,steel}$ entry would be the value of the electricity used by the steel sector. Since it takes steel to make a steel factory, there would also be $m_{steel,steel}$ entries along the diagonal of the matrix. It often is the case that the total, economy-wide, amount of electricity, steel, and other commodities flowing each year through the system are known, but the individual matrix elements are incompletely known. MaxEnt allows inference of the matrix elements (Cho and Judge, 2008).

6.3.4 Food webs and other networks

Matrices somewhat similar to those that arise in economics or in Table 6.2 can also arise in the analysis of ecological food webs. For example, suppose the net primary productivity of each of the species of plants in an ecosystem is known or estimated, and so is the total food intake of each species of herbivore, but the quantity of food that each species of herbivore derives from each plant species is not known. A matrix similar to the two described above can be constructed, with known row and column sums, but unknown matrix elements. MaxEnt has not been applied to estimating the magnitudes of the flows between the species in a food web.

On the other hand, MaxEnt has been used to infer the distribution of numbers of linkages between nodes in an ecological web. Nodes are species and linkages are lines connecting two species that interact with each other. Some nodes are connected to many other nodes, while others are connected sparsely. The metric $\Lambda(l|S_0, L_0)$ in Table 3.1 describes such a distribution. The most commonly studied type of interaction in a network is that in which individuals in one species consume individuals in the second, in which case the network is a food web. As mentioned in Section 3.4.11, Williams (2010), has shown that MaxEnt provides a reasonable description of that distribution. The basic idea is straightforward. If S_0, the number of species, and L_0, the number of trophic linkages, is all we know about a food web, then the mean of $\Lambda(l|S_0, L_0)$ is known to be L_0/S_0. Table 6.1 informs us that knowing just the mean of a distribution results in a MaxEnt prediction of an exponential distribution $\Lambda(l|S_0, L_0) = (1/Z)\cdot\exp(-\lambda\cdot l)$ where Z is given by Eq. 5.13 and λ is determined by applying Eq. 5.14. Conceptually, the procedure is identical to that used in Section 6.2 for fair or unfair die problems.

The MaxEnt distribution of linkages in more specialized webs, such as pollinator–plant networks, host–parasite networks, or any type of mutualistic or competitive set of interactions, can also be analyzed by the same methods. Beyond ecology, the distribution of linkages in internet webs, friendship webs, or any type of network can also be inferred with MaxEnt.

6.3.5 Classical and non-equilibrium thermodynamics and mechanics

In a series of papers, Jaynes (1957a, 1957b, 1963, 1979, 1982), showed that classical thermodynamics can be derived from MaxEnt. I urge readers with some background and interest in physics to read these papers, which in this author's opinion are paragons of brilliance and clarity. At the cutting and controversial edge of MaxEnt applications to thermodynamics is the work of Dewar (2003, 2005), who has attempted to derive a putative law of far-from-equilibrium statistical mechanics called the principle of Maximum Entropy Production (MEP). His approach was to derive from MaxEnt a probability distribution for the number of paths in phase space leading from an initial state to each possible future state. He then shows that the distribution achieves its maximum when the final state is the one in which the maximum possible amount of entropy is produced in the transition from initial to final state. In a physical system, MEP is tantamount to saying that a far-from equilibrium system is irreversibly dissipating energy (through, e.g. friction) as rapidly as possible. We will return to this concept of MEP in Chapter 11, but here I mention that there is a gap in the Dewar derivation, as shown by Grinstein and Linsker (2007), and so, as I write, the status of MEP is unclear.

Equally speculatively, in a recent paper Brissaud (2007) has advanced a derivation of classical mechanics from MaxEnt. One formulation of classical mechanics invokes an extremum principle called the Principle of Least Action (Goldstein 1957); Brissaud presented a parallel derivation starting from MaxEnt.

6.3.6 Macroecology

Several authors (e.g. Shipley et al., 2006; Pueyo et al., 2007; Dewar and Porte, 2008; Azaele et al. 2010) have recently proposed applications of MaxEnt to macroecology, with the goal of predicting limited subsets of the metrics discussed in Chapter 3. I will comment on these efforts in Chapter 10, where I more generally relate the theory advanced in Chapter 7 to a wide range of extant appoaches in theoretical ecology.

6.4 Exercises

Exercise 6.1

With reference to Axiom 4 in Chapter 5, think of an everyday example in which the order in which you acquire two pieces of information influences the final amount of information you possess.

Exercise 6.2

Show that if the only constraining information you have about a six-sided die is that the mean value of results of many tosses exceeds 3.5, then $\lambda < 0$ and the MaxEnt solution for $P(n)$ is an increasing function of n.

Exercise 6.3

Consider a fair six-sided die, which means that the mean value of n is equal to 3.5.

(a) Derive the numerical value of the mean of n^2.
(b) According to Table 6.1, the MaxEnt solution for $p(n)$, under the constraints of knowledge of the mean of n and n^2, is a normal distribution. But we know a fair die has a uniform probability distribution. Show how both statements can be reconciled.

**Exercise 6.4

Suppose you have a probability distribution defined on the integers from 0 to 10, and the only constraining prior information you have is that $P(7) = 0.2$ and the mean is 6. What does MaxEnt lead you to infer?

Part IV

Macroecology and MaxEnt

7

The maximum entropy theory of ecology (METE)

Here I present METE, a comprehensive, parsimonious, and testable theory of the distribution, abundance, and energetics of species across spatial scales. The structure and predictions of the theory are developed in this chapter, with theory evaluation following in Chapter 8.

7.1 The entities and the state variables

Application of the MaxEnt concept to any complex system requires a decision as to the fundamental entities (for example, molecules in the classic thermodynamic example) and specification of a set of state variables (Section 1.3.6). METE predicts probability distributions defined on two kinds of entities, "individuals" and "species," but we do not need to narrowly define either. Thus we take the "species" to mean any well-defined set of groups of "individuals." The groups could be the taxonomic species (as will be assumed here), but they might also be genera, or even trait groups. And "individuals" will be defined in the usual sense of individual organisms, but other choices can be readily accommodated. The only criteria for choosing the entities in a MaxEnt application is that they are unambiguously defined in a manner that allows specification of the numerical values of the state variables and, therefore, as we shall see, of the quantitative constraints on the probability distributions. It is certainly not the case, however, that if MaxEnt yields accurate predictions for one choice of fundamental units, then it will necessarily do so for others.

Recalling the discussion in Section 1.3.6, we choose the state variables for macroecology to be the area, A_0, of the system, the number of species, S_0, from any specified taxonomic category, such as plants or birds, in that area, the total number of individuals, N_0, in all those species, and the total rate of metabolic energy, E_0, consumed by all those individuals. There is no a priori way to justify this or any other choice of state variables; just as in thermodynamics, only the empirical success of theory can justify the choice of the fundamental entities (for example, molecules or individual organisms) and the state variables (pressure, volume, number of moles, or A_0, S_0, N_0, E_0) that theory is built from.

7.2 The structure of METE

Two MaxEnt calculations are at the core of the theory: the first yields all the metrics that describe abundance and energy distributions; and the second describes the spatial-scaling properties of species' distributions. Here we set up the structure of the theory and in Sections 7.3–7.8 derive the actual predicted forms of all the metrics.

7.2.1 Abundance and energy distributions

The core theoretical construct in METE is a joint, conditional probability distribution $R(n,\epsilon|A_0, S_0, N_0, E_0)$. Because of the central role this distribution plays in METE, I will give it a name: the ecosystem structure function. It is defined over the species and individuals (from within any chosen taxonomic category) in A_0, and it describes how individuals are distributed over species and how metabolism is distributed over individuals. R is a discrete distribution in abundance, n, and a continuous distribution in metabolic energy rate ϵ. In particular:

> $R \cdot d\epsilon$ is the probability that if a species is picked from the species' pool, then it has abundance n, and if an individual is picked at random from that species, then its metabolic energy requirement is in the interval $(\epsilon, \epsilon+d\epsilon)$.

Note that ϵ is a rate of energy use, and so it is a measure of power, not energy.

For reasons that will be clear later, we assume that there is a minimum metabolic rate, ϵ_{min}, below which an individual, no matter how small you choose it to be from the pool of N_0 individuals, cannot exist. We define ϵ_{min} to be 1 by choice of energy units. Similarly, when we derive mass distributions, we set the minimum mass = 1, thereby defining the unit of mass. An implication of this choice of units is that if we assume a power-law relationship between metabolic rate and body mass of an individual in the form $\epsilon = c \cdot m^b$, we can take the constant c to equal 1.

The normalization condition on R is then:

$$\sum_{n=1}^{N_0} \int_{\varepsilon=1}^{E_0} d\varepsilon \cdot R(n, \varepsilon|A_0, S_0, N_0, E_0) = 1. \tag{7.1}$$

Two other constraints on the ecosystem structure function result from specification of the three state variables S_0, N_0, and E_0. In particular, these state variables determine two independent ratios: N_0/S_0 is the average abundance per species and E_0/S_0 is the average over species of the total metabolic rate of the individuals within the species. This gives us the following constraint equations on R:

$$\sum_{n=1}^{N_0} \int_{\varepsilon=1}^{E_0} d\varepsilon \cdot n \cdot R(n, \varepsilon|A_0, S_0, N_0, E_0) = \frac{N_0}{S_0}. \tag{7.2}$$

and

$$\sum_{n=1}^{N_0} \int_{\varepsilon=1}^{E_0} d\varepsilon \cdot n \cdot \varepsilon \cdot R(n,\varepsilon|A_0,S_0,N_0,E_0) = \frac{E_0}{S_0}. \qquad (7.3)$$

Area, A_0, is absent from these equation and I drop that conditional variable hereafter. The explicit role of area in our metrics will enter the theory in Section 7.2.2, when the spatial abundance distribution, $\Pi(n|A, n_0, A_0)$, is determined.

The distribution $R(n,\varepsilon)$ can now be obtained by maximizing the information entropy of that distribution:

$$I_R = -\sum_{n=1}^{N_0} \int_{\varepsilon=1}^{E_0} d\varepsilon \cdot R(n,\varepsilon) \cdot \log(R(n,\varepsilon)). \qquad (7.4)$$

subject to the constraints of Eqs 7.1–7.3. By fixing the units used to express the continuous variable, ε, so that the lower limit of integration, corresponding to the minimum possible metabolic rate, is fixed at 1, we eliminate the need for a prior distribution to ensure that the MaxEnt solution for $R(n, \varepsilon)$ is units independent. On grounds of lacking a reason to do otherwise, we invoke no non-uniform prior distribution.

From the distribution R, a number of the metrics introduced in Table 3.1 can be derived. For example, the species–abundance distribution (Section 3.3.6) is given by integrating out the energy variable:

$$\Phi(n|S_0,N_0) = \int_{\varepsilon=1}^{E_0} d\varepsilon \cdot R(n,\varepsilon|S_0,N_0,E_0). \qquad (7.5)$$

Similarly, the community energy distribution (Section 3.3.10) is given by:

$$\Psi(\varepsilon|S_0,N_0) = \frac{S_0}{N_0} \sum_{n=1}^{N_0} n \cdot R(n,\varepsilon|S_0,N_0,E_0). \qquad (7.6)$$

Before delving deeper into the theory and relating other metrics from Tables 3.1 and 3.2 to R, it will be helpful to set up a simple toy example of a community of species so that the meaning of R and the form of these equations, especially the reason for the factor S_0/N_0 in Eq. 7.6, is clearly understood. This is accomplished in Box 7.1. Box 7.2 gives a formal derivation of Eq. 7.6.

From the definition of the ecosystem structure function, $R(n, \varepsilon)$, the intra-specific metabolic rate distribution θ $(\varepsilon|n)$ introduced in Table 3.1 can be readily derived. $\theta(\varepsilon|n)d\varepsilon$ is the probability that an individual has metabolic rate in the interval $\varepsilon, \varepsilon+d\varepsilon$

144 • *The maximum entropy theory of ecology (METE)*

if the species has abundance n, and $\Phi(n)$ is the probability a species has abundance n. Hence, from the fundamental rule $P(A|B)P(B) = P(A,B)$ we can write:

$$\theta(\varepsilon|n) = R(n,\varepsilon)/\Phi(n). \tag{7.7}$$

We shall see shortly that from the metrics Θ and Φ, the other energy metrics in Table 3.2 can be derived. And from those energy metrics, the mass metrics in Table 3.3 can be derived, if a mass–metabolism relationship, such as the metabolic scaling law $\epsilon = m^{3/4}$, is assumed.

Box 7.1 A toy community

Consider an ecosystem with $S_0 = 4$, $N_0 = 16$, and $E_0 = 64$. For convenience, we take the metabolic rate, ϵ, of individuals to be a discrete variable, and so $R(n,\epsilon)$ is now a discrete distribution in both variables. The actual assignments of abundance and metabolic rates are shown in Table 7.1:

Table 7.1 The abundances of the four species, and the metabolic rates of the 16 individuals in a simple community with total metabolic rate of 64.

Species	Abundance	Metabolic rates of individuals	Total metabolic rate per species
A	1	23	23
B	1	9	9
C	2	6, 9	15
D	12	1,1,1,1,1,1,1,1,2,2,2,3	17
$S_0 = 4$	$N_0 = 16$	$E_0 = 64$	$E_0 = 64$

Here are the values of the $R(n,\epsilon)$:
$R(12,1) =$ (the probability that if a species is picked from the species pool it has abundance $= 12$) $\times$ (the probability that if an individual is picked from such a species it has metabolic rate $= 1$) $= (1/4)\cdot(8/12) = 8/48$.
The other non-zero values or R are similarly obtained:

$$R(12,2) = (1/4)\cdot(3/12) = 3/48$$
$$R(12,3) = (1/4)\cdot(1/12) = 1/48$$
$$R(2,6) = (1/4)\cdot(1/2) = 6/48$$
$$R(2,9) = (1/4)\cdot(1/2) = 6/48$$
$$R(1,9) = (1/4)\cdot(1) = 12/48$$
$$R(1,23) = (1/4)\cdot(1) = 12/48.$$

Note that if n and ϵ are summed over, R is properly normalized to one. Now we can work out the abundance distribution $\Phi(n)$ by summing over ϵ:

$$\Phi(12) = R(12,1) + R(12,2) + R(12,3) = 1/4$$
$$\Phi(2) = R(2,6) + R(2,9) = 1/4$$
$$\Phi(1) = R(1,9) + R(1,23) = \tfrac{1}{2}.$$

Let's see if the energy distribution as expressed in Eq. 7.6 comes out right. According to that equation:

Box 7.1 *(Cont.)*

$$\Psi(1) = (4/16)\cdot(12)\cdot R(12,1) = 1/2.$$

This makes sense; one half of all the individuals in the community have $\epsilon = 1$. The other values of the distribution are:

$$\Psi(2) = (4/16)\cdot(12)\cdot R(12,2) = 3/16$$
$$\Psi(3) = (4/16)\cdot(12)\cdot R(12,3) = 1/16$$
$$\Psi(6) = (4/16)\cdot(2)\cdot R(2,6) = 1/16$$
$$\Psi(9) = (4/16)\cdot[2\cdot R(2,9) + 1\cdot R(1,9)] = 2/16$$
$$\Psi(23) = (4/16)\cdot(1)\cdot R(1,23) = 1/16.$$

Note that both the abundance and energy distributions come out properly normalized. You can verify that both constraint equations (Eqs 7.2 and 7.3) also come out right. Finally, we can derive the values of the conditional species-level metabolic rate distribution $\theta(\epsilon|n)$. From Eq. 7.7, we obtain, for example:

$$\theta(1|12) = R(12,1)/\Phi(1) = (1/6)/(1/4) = 2/3.$$

Indeed, 2/3 of the individuals in the species with $n_0 = 12$, have $\epsilon = 1$.

Box 7.2 Justification of Eq. 7.6

The fraction of species with abundance n is estimated by $\Phi(n)$, and so the expected total abundance of species with abundance n is $nS_0\Phi(n)$. If an individual is picked at random from the individuals' pool (N_0), as opposed to the species' pool, then the probability that the individual belongs to a species with abundance n is given by $nS_0\Phi(n)/N_0$. Similarly, $\Psi(\epsilon)d\epsilon$ is the probability that an individual picked from the entire individuals' pool (N_0) has a metabolic energy requirement between ϵ and $\epsilon + d\epsilon$. Now recall that $\Theta(\epsilon|n)$ equals the conditional probability that the metabolic rate of an individual is between ϵ and $\epsilon + d\epsilon$, if the individual is from a species with abundance n. Then from the general relationship that:

$$P(x) = \sum_y P(x|y)\cdot P(y), \quad (7.8)$$

we have:

$$\Psi(\varepsilon)d\varepsilon = \sum_{n=1}^{N_0} n\cdot\frac{S_0}{N_0}\cdot\Phi(n)\cdot\Theta(\varepsilon|n). \quad (7.9)$$

Using Eq. 7.7, we immediately obtain Eq. 7.6.

7.2.2 Species-level spatial distributions across multiple scales

The focus here is on the metric $\Pi(n|A, n_0, A_0)$. We shall see that there are two ways, each a-priori equally plausible, in which we can use MaxEnt to derive this metric at various spatial scales; only comparison with data can inform us of which is best. At any scale, A, we have the normalization equation

$$\sum_{n=0}^{n_0} \Pi(n|A, n_0, A_0) = 1. \tag{7.10}$$

Because the mean value of n, across cells of area A, is given by:

$$\bar{n} = n_0 A / A_0, \tag{7.11}$$

we also have the constraint equation:

$$\sum_{n=0}^{n_0} n \cdot \Pi(n|A, n_0, A_0) = \frac{n_0 A}{A_0}. \tag{7.12}$$

To obtain $\Pi(n)$, we then maximize the information entropy: $-\Sigma_n \Pi(n) \log(\Pi(n))$ subject to the constraints of Eqs 7.10 and 7.12.

7.3 Solutions: R(n, ε) and the metrics derived from it

Using the methods spelled out in Box 5.1, maximization of information entropy I_R in Eq. 7.4 yields:

$$R(n, \varepsilon) = \frac{1}{Z} e^{-\lambda_1 n} e^{-\lambda_2 n \varepsilon}. \tag{7.13}$$

The partition function, Z, is given by:

$$Z = \sum_{n=1}^{N_0} \int_{\varepsilon=1}^{E_0} d\varepsilon \cdot e^{-\lambda_1 n} e^{-\lambda_2 n \varepsilon}, \tag{7.14}$$

and the two Lagrange multipliers, λ_1 and λ_2, are given by application of Eq. 5.14, with $f_1 = n$, $<f_1> = N_0/S_0$, and $f_2 = n\epsilon$, $<f_2> = E_0/S_0$. For convenience later, we now define:

$$\beta = \lambda_1 + \lambda_2 \tag{7.15a}$$

$$\sigma = \lambda_1 + E_0 \lambda_2 \tag{7.15b}$$

$$\gamma = \lambda_1 + \varepsilon \lambda_2 \tag{7.15c}$$

Box 7.3 derives the exact solutions to the MaxEnt problem posed above. Because it is quite complicated, a huge simplification that is valid under nearly every situation encountered in real ecosystems will be derived in Box 7.4. Boxes 7.3 and 7.4 are lengthy and technical, but following them the key results, the predicted macroecological metrics, are highlighted.

Box 7.3 Exact solution for the ecosystem structure function $R(n,\epsilon)$

The integrals in Eqs 7.2 and 7.3 can be done exactly (Exercise 7.1), yielding:

$$\frac{N_0}{S_0} = \frac{\sum_{n=1}^{N_0}(e^{-\beta n} - e^{-\sigma n})}{Z \cdot \lambda_2} \tag{7.16}$$

and

$$\frac{E_0}{S_0} = \sum_{n=1}^{N_0} \frac{(e^{-\beta n} - E_0 e^{-\sigma n})}{Z \cdot \lambda_2} + \sum_{n=1}^{N_0} \frac{(e^{-\beta n} - e^{-\sigma n})}{n Z \lambda_2^2}. \tag{7.17}$$

And the integral in Eq. 7.14, which enforces the normalization condition Eq. 7.1, yields:

$$Z = \sum_{n=1}^{N_0} \frac{(e^{-\beta n} - e^{-\sigma n})}{\lambda_2 \cdot n}. \tag{7.18}$$

Substituting $Z \cdot \lambda_2$ from Eq. 7.18 into Eqs 7.16 and 7.17, we get (Exercises 7.2, 7.3):

$$\frac{N_0}{S_0} = \frac{\sum_{n=1}^{N_0}(e^{-\beta n} - e^{-\sigma n})}{\sum_{n=1}^{N_0} \frac{(e^{-\beta n} - e^{-\sigma n})}{n}} \tag{7.19}$$

and

$$\frac{E_0}{S_0} = \frac{\sum_{n=1}^{N_0}(e^{-\beta n} - E_0 e^{-\sigma n})}{\sum_{n=1}^{N_0} \frac{(e^{-\beta n} - e^{-\sigma n})}{n}} + \frac{1}{\lambda_2}. \tag{7.20}$$

Equations 7.19 and 7.20 are two equations with two unknowns, β and σ, and therefore, using Eqs 7.15a and b, they determine the two Lagrange multipliers, λ_1 and λ_2, as function of the state variables S_0, N_0, and E_0.

Box 7.3 (*Cont.*)

Although these equations are complicated and have to be solved numerically, for realistic values of the state variables, we will see that extremely accurate simplifications of these equations are possible. First, however, we can derive some exact results. Recalling that:

$$\sum_{n=1}^{N_0} e^{-\beta n} = \frac{e^{-\beta} - e^{-\beta(N_0+1)}}{1 - e^{-\beta}}, \quad (7.21)$$

we can combine Eqs 7.18 and 7.19 to obtain the partition function, Z:

$$Z = \frac{S_0}{\lambda_2 N_0} \sum_{n=1}^{N_0} (e^{-\beta n} - e^{-\sigma n}) = (\frac{S_0}{\lambda_2 N_0})(\frac{e^{-\beta} - e^{-\beta(N_0+1)}}{1 - e^{-\beta}} - \frac{e^{-\sigma} - e^{-\sigma(N_0+1)}}{1 - e^{-\sigma}}). \quad (7.22)$$

From Eq. 7.5, we can now derive the species abundance distribution $\Phi(n)$:

$$\Phi(n) = \Phi(n|S_0, N_0) = \int_{\varepsilon=1}^{E_0} d\varepsilon \cdot R(n, \varepsilon | S_0, N_0, E_0) = \frac{e^{-\lambda_1 n}}{\lambda_2 Z n} (e^{-\lambda_2 n} - e^{-\lambda_2 n E_0})$$

$$= \frac{e^{-\beta n} - e^{-\sigma n}}{\lambda_2 Z n}, \quad (7.23)$$

where $Z\lambda_2$ can be obtained from Eq. 7.22.

Likewise, from Eq. 7.6, we obtain the energy distribution $\Psi(\epsilon)$:

$$\Psi(\varepsilon) = \frac{S_0}{N_0} \sum_{n=1}^{N_0} n \cdot R(n, \varepsilon | S_0, N_0, E_0) = \frac{S_0}{N_0 Z} [\frac{e^{-\gamma}}{(1-e^{-\gamma})^2} - (\frac{e^{-\gamma N_0}}{1-e^{-\gamma}})(N_0 + \frac{e^{-\gamma}}{1-e^{-\gamma}})], \quad (7.24)$$

where γ depends on ϵ as given in Eq. 7.15c.

Next, from Eqs 7.13 and 7.23 we obtain the intra-specific metabolic energy probability density:

$$\Theta(\varepsilon|n) = \frac{R}{\Phi} = \frac{n\lambda_2 e^{-\lambda_2 n\varepsilon}}{e^{-\lambda_2 n} - e^{-\lambda_2 n E_0}}. \quad (7.25)$$

So far, our results are exact. The Lagrange multiplier, λ_1 ad λ_2, can be solved for numerically, using Eqs 7.15.a,b, 7.19 and 7.20 (see Table 7.2 for examples) and the distribution $R(n, \epsilon)$ is uniquely determined from the values of the state variables and the MaxEnt criterion. And from Φ, Ψ, and Θ, all the energy and mass metrics in Table 3.2 can be derived.

Box 7.4 Good approximations to the exact solutions

To derive a simple expression for λ_2 we note that we can ignore the terms in Eqs 7.19 and 7.20 with $e^{-\sigma n}$. This is because for realistic values of the state variables (see Table 7.2), $\beta < 1$, σ is $\geq S_0$, and λ_2 is $\sim S_0/E_0$. As a consequence, terms of order $e^{-\sigma}$ are very small compared to terms of order $e^{-\beta}$ and terms of order $E_0 e^{-\sigma}$ are very small compared with terms of order $1/\lambda_2$.

Dropping the terms in Eqs 7.19 and 7.20 with $e^{-\sigma n}$, we can combine the equations and obtain (Exercise 7.4):

$$\lambda_2 \approx \frac{S_0}{E_0 - N_0}. \tag{7.26}$$

Note that λ_2 is always positive because N_0 can never exceed E_0 (see Exercise 7.5). Dropping the terms with $e^{-\sigma}$ in Eq. 7.19, we obtain:

$$\frac{N_0}{S_0} \approx \frac{\sum_{n=1}^{N_0} e^{-\beta n}}{\sum_{n=1}^{N_0} \frac{e^{-\beta n}}{n}}. \tag{7.27}$$

A further simplification can be obtained if an approximation to the summation in the denominator of Eq. 7.19 is made. In particular, if $\beta N_0 \gg 1$ and $\beta \ll 1$, which we will see follows from the assumption that $S_0 \gg 1$, then the summation is well approximated by (see Abramowitz and Stegen, 1972):

$$\sum_{n=1}^{N_0} \frac{e^{-\beta n}}{n} \approx \log\left(\frac{1}{\beta}\right). \tag{7.28}$$

The summation in the numerator of Eq. 7.19 can be done exactly (see Eq. 7.21), and so Eq. 7.27 becomes:

$$\frac{S_0}{N_0} \approx \frac{1 - e^{-\beta}}{e^{-\beta} - e^{-\beta(N_0+1)}} \cdot \log\left(\frac{1}{\beta}\right). \tag{7.29}$$

Furthermore, if $\beta \ll 1$, then this can be further simplified to:

$$S_0/N_0 = \beta \log(1/\beta). \tag{7.30}$$

Moreover, using Eq. 7.28, the expression in Eq. 7.22 for Z can now be approximated by:

$$Z \approx \left(\frac{S_0}{\lambda_2 N_0}\right)\left(\frac{e^{-\beta} - e^{-\beta(N_0+1)}}{1 - e^{-\beta}}\right) \approx \frac{\log\left(\frac{1}{\beta}\right)}{\lambda_2}. \tag{7.31}$$

Although Eq. 7.30 cannot be solved for β analytically, it is much easier to solve numerically than is Eq. 7.19. Throughout the rest of this book, we will make use of this approximation to Eq. 7.19 when the values of the state variables justify doing so.

Box 7.4 (*Cont.*)

Table 7.2 shows numerical solutions for β and σ, as well as for the two Lagrange multipliers, for a wide range of state variables. There are several things worth noting in the Table. First, σ is indeed $\geq S_0$ as mentioned above, Second, the solutions to the approximate Eq. 7.30 tend to agree most closely with the solutions to the exact Eq. 7.19 for those cases with highest S_0. This is because the joint conditions that make Eq. 7.30 a good approximation, $\beta \ll 1$ and $\beta N_0 \gg 1$, are increasingly valid for large S_0. Third, we note that $\gamma N_0 > \beta N_0 \gg 1$ for large S_0, which will be relevant below when we simplify the expression for $\Psi(\epsilon)$.

Finally, we note that in real ecosystems, we expect $\lambda_2 \ll 1$. To see why, first note that from the definition of the species–abundance distribution $\Phi(n)$, $\Phi(n)$, $N_0 = \sum_{n=1}^{N_0} n \cdot \Phi(n)$ and $E_0 = \sum_{n=1}^{N_0} \varepsilon(n) \cdot n \cdot \Phi(n)$. Then, because we expect that the individuals in many species will have metabolic rates much greater than that of the individual with the minimum rate, $\epsilon = 1$, it follows that E_0 will likely greatly exceed N_0. Because S_0/N_0 is generally much less than 1 in most ecosystems, it then follows from Eq. 7.26 that $\lambda_2 \ll 1$.

Fortunately, the rather complicated equations in Box 7.3 can be greatly simplified through a series of approximations that are generally well-justified. Box 7.4 presents these simplifications. The condition under which the results derived in Box 7.4 are valid is that $e^{-S_0} \ll 1$. For most combinations of the other state variables, validity of the approximations is assured if $S_0 > 4$.

Assuming $e^{-S_0} \ll 1$, Eqs 7.23–7.25 in Box 7.3, along with the results in Box 7.4, lead directly to the following approximate forms for the distributions $\Phi(n)$ and $\Psi(\epsilon)$ and $\Theta(\epsilon|n)$. We write these distributions now with the conditional variables made explicit:

$$\Phi(n|S_0, N_0) \approx \frac{1}{\log(\beta^{-1})} \cdot \frac{e^{-\beta n}}{n} \tag{7.32}$$

and

$$\Psi(\varepsilon|S_0, N_0, E_0) \approx \lambda_2 \cdot \beta \cdot \frac{e^{-\gamma}}{(1 - e^{-\gamma})^2} \tag{7.33}$$

and

$$\Theta(\varepsilon|n, S_0, N_0, E_0) \approx \lambda_2 n e^{-\lambda_2 n(\varepsilon - 1)}. \tag{7.34}$$

The value of β is given by either the more exact Eq. 7.27 or the approximate Eq. 7.30. Referring to Eq. 7.15, The value of γ is given in terms of the values of β and λ_2, and λ_2 is given by Eq. 7.26. Exercise 7.6 asks you to verify the validity of Eqs 7.32–7.34.

Comparing these metrics to empirical data is best accomplished with the cumulative distribution functions, or equivalently the rank–abundance and rank–

Table 7.2 Numerical solutions for the Lagrange multipliers in METE. The values of the state variables, S_0, N_0, E_0, are chosen to span a wide range of conditions that, as we shall see in Chapter 8, are within the range found in many ecosystems for which there are census data that can be used for testing macroecological theory. We have chosen values of N_0 that are four different multiples of S_0, and values of E_0 that are two different multiples of N_0 to more clearly distinguish the influence of varying the ratios of state variables from varying their magnitudes.

S_0	N_0	E_0	β from Eq. 7.27	β from Eq. 7.30	σ	λ_1	λ_2	βN_0
4	$2^2 S_0$	$2^2 N_0$	0.0459	0.116	5.4	−0.037	0.083	0.74
4	$2^4 S_0$	$2^2 N_0$	−0.00884	0.0148	5.3	−0.030	0.021	−0.56
4	$2^8 S_0$	$2^2 N_0$	−0.00161	0.000516	5.3	−0.0029	0.0013	−1.6
4	$2^{12} S_0$	$2^2 N_0$	−0.000135	0.0000229	5.3	−0.00022	0.000081	−2.3
4	$2^2 S_0$	$2^{10} N_0$	0.0459	0.116	4.0	0.046	0.00024	0.74
4	$2^4 S_0$	$2^{10} N_0$	−0.00884	0.0148	4.0	−0.0089	0.000061	−0.6
4	$2^8 S_0$	$2^{10} N_0$	−0.00161	0.000516	4.0	−0.0016	3.82E−06	−1.6
4	$2^{12} S_0$	$2^{10} N_0$	−0.000135	0.0000229	4.0	−0.00014	2.39E−07	−2.3
16	$2^2 S_0$	$2^2 N_0$	0.101	0.116	21.4	0.018	0.083	6.4
16	$2^4 S_0$	$2^2 N_0$	0.0142	0.0148	21.3	−0.0066	0.021	3.6
16	$2^8 S_0$	$2^2 N_0$	0.000413	0.000516	21.3	−0.00089	0.0013	1.7
16	$2^{12} S_0$	$2^2 N_0$	0.0000122	0.0000229	21.3	−0.000069	0.000081	0.79
16	$2^2 S_0$	$2^{10} N_0$	0.101	0.116	16.1	0.10	0.00024	6.4
16	$2^4 S_0$	$2^{10} N_0$	0.0142	0.0148	16.0	0.014	0.000061	3.6
16	$2^8 S_0$	$2^{10} N_0$	0.000413	0.000516	16.0	0.00041	3.82E−06	1.7
16	$2^{12} S_0$	$2^{10} N_0$	0.0000122	0.0000229	16.0	0.000012	2.39E−07	0.79
64	$2^2 S_0$	$2^2 N_0$	0.102	0.116	85.4	0.018	0.083	26
64	$2^4 S_0$	$2^2 N_0$	0.0147	0.0148	85.3	−0.0061	0.021	15
64	$2^8 S_0$	$2^2 N_0$	0.000516	0.000516	85.3	−0.00079	0.0013	8.5
64	$2^{12} S_0$	$2^2 N_0$	0.0000228	0.0000229	85.3	−0.000059	0.000081	6.0
64	$2^2 S_0$	$2^{10} N_0$	0.102	0.116	64.2	0.10	0.00024	26
64	$2^4 S_0$	$2^{10} N_0$	0.0147	0.0148	64.1	0.015	0.000062	15
64	$2^8 S_0$	$2^{10} N_0$	0.000516	0.000516	64.1	0.00051	3.82E−06	8.5
64	$2^{12} S_0$	$2^{10} N_0$	0.0000228	0.0000229	64.1	0.000023	2.39E−07	6.0
256	$2^2 S_0$	$2^2 N_0$	0.102	0.116	341.4	0.018	0.083	102
256	$2^4 S_0$	$2^2 N_0$	0.0147	0.0148	341.3	−0.0061	0.021	61
256	$2^8 S_0$	$2^2 N_0$	0.000516	0.000516	341.3	−0.00079	0.0013	34
256	$2^{12} S_0$	$2^2 N_0$	0.0000228	0.0000229	341.3	−0.000059	0.000081	24
256	$2^2 S_0$	$2^{10} N_0$	0.102	0.116	256.4	0.10	0.00024	102
256	$2^4 S_0$	$2^{10} N_0$	0.0147	0.0148	256.3	0.015	0.000062	61
256	$2^8 S_0$	$2^{10} N_0$	0.000516	0.000516	256.3	0.00051	3.82E−06	34
256	$2^{12} S_0$	$2^{10} N_0$	0.0000228	0.0000229	256.3	0.000023	2.39E−07	24

metabolism functions derived from them. In the following subsection, we derive closed-form expressions for the rank–energy functions from Eqs 7.33 and 7.34 and, discuss the rank–abundance relationship derived from Eq. 7.32. It is not possible to derive a closed-form rank–abundance relationship from Eq. 7.32, but a useful and simple expression for the predicted number of rare species can be derived, as shown in Section 7.3.2.

7.3.1 Rank distributions for $\Psi(\varepsilon)$, $\Theta(\varepsilon)$, and $\Phi(n)$

Consider, first, the distribution of metabolic rates over all the individuals in the community. For this we turn to $\Psi(\epsilon)$, given in Eq. 7.33. Referring to Box 3.4, which shows how to relate rank–variable relationships to probability distributions, we can derive the rank–metabolism relationship corresponding to $\Psi(\epsilon)$. We will associate with rank $r = 1$ the individual with the highest metabolic rate and with rank $r = N_0$ value the individual with the lowest rate. Recalling that metabolic rate units are chosen so that the minimum possible rate is one, we have:

$$\int_1^{\varepsilon_\Psi(r)} d\varepsilon \Psi(\varepsilon) = \frac{1}{N_0} \int_{r-1/2}^{N_0} dr = 1 - \frac{r - 1/2}{N_0}. \tag{7.35}$$

The factor of ½ in this expression warrants some explanation. It is reasonable to assume that the probability of an individual having $\epsilon > \epsilon(r = 1)$ can be taken to be $1/(2N_0)$. The factor of 2 in this expression is there because, on a graph in which metabolic rate is plotted on the horizontal axis, and along which each individual gets an equal share, $1/N_0$, of probability, the most energetic individual should have a metabolic rate that is intermediate between the maximum possible rate it can have (E_0) and a rate that is roughly midway between its actual rate and that of the individual with the second-highest rate. Hence, roughly half of its share of probability ($1/N_0$) should lie with values of $\epsilon > \epsilon_{\max}$. Fortunately the rank distribution is not very sensitive to the factor of $1/2$ in the expression ½ N_0.

Using Eq. 7.33, the remaining integral in Eq. 7.35 can be carried out by making a change of variable: $x= \exp(-\gamma)$, where $\gamma = \lambda_1 + \lambda_2 \epsilon$. Dropping terms of order β^2, this results in (see Exercise 7.7):

$$\int_1^{\varepsilon_\Psi(r)} d\varepsilon \Psi(\varepsilon) \approx 1 - \frac{\beta}{1 - e^{-(\lambda_1 + \lambda_2 \varepsilon_\Psi(r))}}. \tag{7.36}$$

Equating this expression to the right-hand side in Eq. 7.35 allows us to write:

$$\varepsilon_\Psi(r) = \frac{1}{\lambda_2} \log\left(\frac{\beta N_0 + r - \frac{1}{2}}{r - \frac{1}{2}}\right) - \frac{\lambda_1}{\lambda_2}. \tag{7.37}$$

The subscript on ϵ is there to remind us that this is the rank–metabolism relationship for all the individuals in the community. This expression can then be readily compared with rank–metabolism data.

A similar approach yields the rank–metabolism rate for the individuals in a species. For this, we turn to the distribution $\Theta(\epsilon|n)$ in Eq. 7.34. Because it is defined over the individuals within a species of abundance n, the rank values will extend from 1 (most energetic individual) to n (least energetic individual). The result is:

$$\varepsilon_\Theta(r,n) = 1 + \frac{1}{\lambda_2 n}\log\left(\frac{n}{r-\frac{1}{2}}\right). \tag{7.38}$$

The rank–abundance relationship corresponding to the species–abundance distribution given in Eq. 7.32 cannot be found in closed form because a closed-form cumulative summation over the logseries distribution does not exist. To compare the rank–abundance relationship obtained from Eq. 7.32 with observed abundances, we can evaluate the cumulative summation numerically, however. A simple way to do this with a spreadsheet is to enter in column A, integers, n, from 1 to N_0. Column B then lists the values of $\Phi(n)$ obtained by numerical evaluation of Eq. 7.32. Each entry in column C is S_0 times the cumulative sum of Φ from abundance 1 to abundance n; because Φ is normalized to 1, the last entry in column C, corresponding to $n = N_0$, will be S_0. The expected values of the abundances of the S_0 species can now be read off from the values in column C. For example, the expected abundance of the most abundant species will be that value of n such that the cumulative sum in column C equals $S_0 - \frac{1}{2}$. That would be the abundance of the rank = 1 species. The next highest abundance (rank = 2) will correspond to a cumulative sum of $S_0 - 3/2$, and so on. Typically, there will be more than 1 species predicted to have $n = 1$, as will be evident from an entry in column C in the row corresponding to $n = 1$ that is greater than $\frac{1}{2}$. If, for example, the entry value is 3.5, then that would indicate that the expected number of species with $n = 1$ is approximately 4. If the entry were, say, 4, then there would be between 4 and 5 such species expected.

If, instead of using this rather simple approach, one were to use software that carries out a Monte Carlo sampling from the probability distribution, Φ, the results would be nearly the same. If 1000 Monte Carlo runs are carried out from a distribution that leads to a column-C entry of 4 in the $n = 1$ row, then the average yield of species with $n = 1$ would be 4.

7.3.2 Implications: extreme values of n and ε

From these rank distributions, some useful information can be derived. First, note that the metabolic rate of the most energetic individual can be written down immediately:

$$\varepsilon_{\Psi,max} = \varepsilon_\Psi(r=1) \approx \frac{1}{\lambda_2}\log(2\beta N_0). \tag{7.39}$$

It is also possible to predict the number of very rare species from knowledge of the state variables S_0 and N_0. The reason is that the state variables determine β, and β determines the function $\Phi(n)$. If we sample repeatedly from the discrete distribution $\Phi(n)$, we expect that the number of species with abundance n will be approximately

$S_0 \cdot \Phi(n)$. Suppose we now ask: how many species in the community have no more than, say, 10 individuals. This choice of 10 is simply an illustrative judgment about what constitutes rarity and thus endangerment. Calling that number $S(n \leq 10)$, we derive from Eq. 7.32:

$$S(n \leq 10) = \frac{S_0}{\log(1/\beta)} \sum_{n=1}^{10} \frac{e^{-\beta n}}{n} \approx \frac{S_0}{\log(1/\beta)} \sum_{n=1}^{10} \frac{1}{n} \approx \frac{2.93 \cdot S_0}{\log(1/\beta)} \approx 2.93 \cdot \beta \cdot N_0. \tag{7.40}$$

Approximating the term $e^{-\beta n}$ in Eq. 7.40 with 1 is justified provided our choice of a cutoff for rarity is $\ll 1/\beta$. For later reference, note that the predicted number of singleton species (species with $n = 1$) is just βN_0. The sum of the inverses of the first m integers is approximately $\gamma + \log(m)$; Euler's constant, $\gamma \approx 0.577$.

At the other extreme of abundance, the theory predicts the abundance, $n_{\max}$, of the most common species in the plot. We can write:

$$\sum_{n=1}^{n_{\max}} \Phi(n) \approx \sum_{n=1}^{n_{\max}} \frac{e^{-\beta \cdot n}}{n \cdot \log(1/\beta)} \approx 1 - \frac{1}{2S_0}. \tag{7.41}$$

The term on the right-hand side in this expression can be understood by the same argument given following Eq. 7.35. If, as is usually the case, $\beta n_{\max}$ is not $\ll 1$, the solution to Eq. 7.41 has to be determined numerically.

As shown above, the number of species with a single individual is predicted to be simply βN_0. From the representative values of the state variables in Table 7.2, it can be seen that, even if $N_0 \gg S_0$, the number of rare species can be substantial. For example, with $S_0 = 256$, $N_0 \sim 10^6$, nearly 10% of the species are predicted to be singletons. This might seem strange. After all, a species represented by only 1 individual should have a poor chance of existing very long. We have to recognize, however, the nature of the prediction. We have taken a prescribed area, presumably within some larger biome, and examined a plot of area A_0, with state variables S_0, N_0. If we were to double the area, we would double N_0 and increase S_0 more modestly. Referring to Table 7.2 again, this implies that $\beta \cdot N_0$ increases but by less than a doubling, because β decreases. Hence the fraction of species that are singletons, $\beta \cdot N_0 / S_0 \approx 1/\log(1/\beta)$ decreases. As one increases the spatial scale of the census, species that were singletons at small scale will tend to have more than 1 individual at larger scale, but new singleton species will be seen. This would continue as scale gets ever larger.

A problem does arise as scale increases and one approaches the entire area of a biome. At that scale, singleton species are still predicted to occur, even though the fraction of such species has decreased. This could be interpreted in one of three ways. It might mean that the theory cannot be believed at such large scales. Or it might mean that these predicted singletons are species on their way out of existence

Solutions: R(n, ϵ) and the metrics derived from it • 155

in that biome, to be replaced by other species that will become singletons for a while. Or, it might mean that we should interpret "individuals" within the context of the theory as mating pairs and that extreme rarity can persist longer than we might have thought. Further study is needed to resolve this.

7.3.3 Predicted forms of other energy and mass metrics

From Eq. 7.34, we can now obtain the predicted form of the energy metric $\bar{\epsilon}(n|\ldots)$ in Table 3.2, the dependence on abundance of the average metabolic rate of the individuals in a species:

$$\bar{\epsilon}(n|S_0, N_0, E_0) = \int d\epsilon \cdot \epsilon \cdot \Theta(\epsilon|n) \approx 1 + \frac{1}{n\lambda_2}. \tag{7.42}$$

Now we can see what the conditional variables are that $\bar{\epsilon}(n)$ depends upon: precisely the state variables in the combination (Eq. 7.26) that determine λ_2.

From Eq. 7.42, we can derive the form of $v(\bar{\epsilon}|\ldots)$ in Table 3.2, the distribution of averaged metabolic rates across all species. First we note that from Eq. 7.42:

$$\frac{d\bar{\epsilon}}{dn} \approx \frac{-1}{n^2 \lambda_2}. \tag{7.43}$$

Hence, using Eq. 7.32:

$$v(\bar{\epsilon}|S_0, N_0, E_0) = \Phi(n(\bar{\epsilon}))|(\frac{d\bar{\epsilon}}{dn})^{-1}| \approx \frac{1}{\log(\frac{1}{\beta})} \cdot \frac{e^{-\beta n(\bar{\epsilon})}}{n(\bar{\epsilon})} \cdot n^2(\bar{\epsilon})\lambda_2$$

$$= \frac{1}{\log(\frac{1}{\beta})} \cdot \lambda_2 n(\bar{\epsilon}) e^{-\beta n(\bar{\epsilon})} \tag{7.44}$$

We see that the conditional variables determining the distribution of average metabolic rates across species are the state variables in the combinations that determine λ_2 and β.

To see more clearly the shape of this energy distribution, we use Eq. 7.42 to write:

$$n(\bar{\epsilon}) \approx \frac{1}{\lambda_2(\bar{\epsilon} - 1)}. \tag{7.45}$$

Substituting this into Eq. 7.44 gives us:

$$v(\bar{\epsilon}|S_0, N_0, E_0) \approx \frac{1}{\log(\frac{1}{\beta})} \frac{e^{-\frac{\beta}{\lambda_2(\bar{\epsilon}-1)}}}{(\bar{\epsilon} - 1)}. \tag{7.46}$$

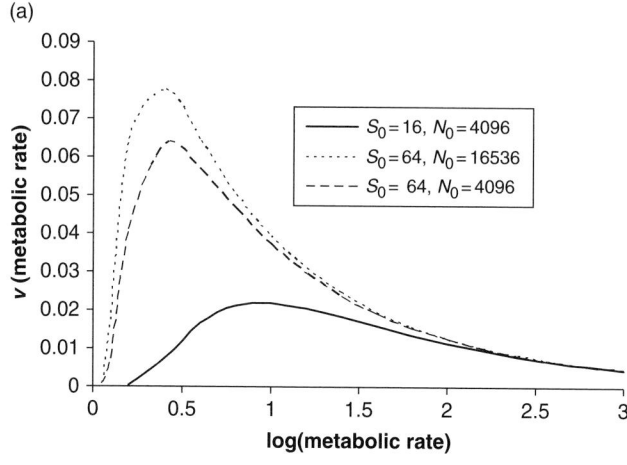

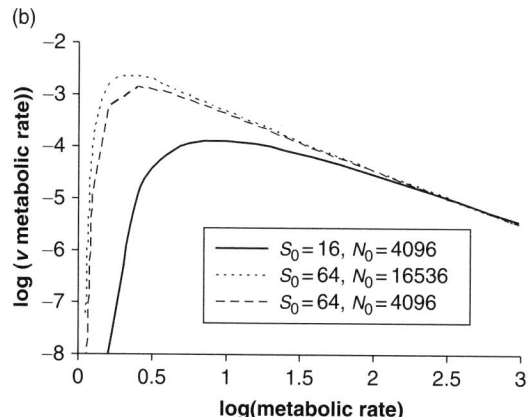

Figure 7.1 (a) The predicted function $v(\bar{\epsilon})$ in Eq. 7.37 plotted against log ($\bar{\epsilon}$) for three sets of values of the state variables. For each combination of S_0 and N_0, the total metabolic rate E_0 is 65,536. (b) Same as (a), but the logarithm of the distribution is plotted.

The term β/λ_2 in the exponent will be $\gg 1$, if $E_0 \gg N_0 \gg S_0$ (see Table 7.1), which is often the case. Equation 7.46 predicts a unimodal (ti.e. hump-shaped) distribution, rising rapidly at small $\epsilon - 1$ to a value peaking at $\epsilon = \beta/\lambda_2$, and then, for $\epsilon \gg \beta/\lambda_2$, declining slowly, as $1/(\epsilon - 1)$. Examples of this behavior are shown in Figure 7.1, where the distributions are plotted against log ($\bar{\epsilon}$) for several choices of the state variables. Regardless of the values of the state variables, and therefore of β/λ_2, the distributions plotted against log ($\bar{\epsilon}$) exhibit an extended tail to the right of the mode. Despite appearances that result from plotting the distribution against the logarithm of $\bar{\epsilon}$, the distributions are each normalized to 1.

To compare against data the predicted distribution, across species, of the metabolic rates averaged over individuals within species, the rank versus metabolic-rate distribution is needed. As with Φ, an analytically tractable cumulative integral over $\bar{\epsilon}$ of $\nu(\bar{\epsilon})$ is not possible. Nevertheless, the same numerical procedure described in Section 7.3.1 for evaluating the predicted rank–abundance distribution can be used to work out the predicted rank versus metabolic-rate distribution. $\nu(\bar{\epsilon})$ will be properly normalized, if the values of $\bar{\epsilon}$ are allowed to range between $\bar{\epsilon}(n_{max})$ and $\bar{\epsilon}(n_{min})$, where n_{max} and n_{min} are the limits of the range of n used in normalizing $\Phi(n)$.

If a metabolic scaling relationship relates mass and metabolic rate, in the form $\epsilon = m^b$, then the mass metrics in Table 3.2, $\rho(m|n_0,\ldots)$, $\Xi(m|\ldots)$, $\bar{m}(n_0|\ldots)$, and $\mu(\bar{m}|\ldots)$ can all be derived using the formulas in Section 3.3.11 and the methods presented in Box 3.2.

Table 7.3 presents the predicted functional forms for all the metrics derived from $R(n,\epsilon)$, evaluated under the assumption that the state variables are in the range where the approximations made in Box 7.3 are valid; for the mass metrics $\Xi(m)$ and $\rho(m)$ in the table, no assumptions are made about the dependence of ϵ on m, but for $\bar{m}(n)$ and $\mu(\bar{m})$, the results in the table are based on the additional assumption that some metabolic scaling power-law holds, so that $\epsilon = m^b$ and $d\epsilon/dm = b \cdot m^{(b-1)}$. The implications of these results for the energy-equivalence principle and mass–abundance relationships will be discussed in Section 7.5.

7.4 Solutions: $\Pi(n)$ and the metrics derived from it

In addition to the normalization condition, Eq. 7.10, a single additional constraint can be imposed on the species-level spatial abundance distribution $\Pi(n|A, n_0, A_0)$. This constraint, expressed in Eq. 7.12, can be thought of as arising in either of two ways. Either the abundance, n_0 is known in advance because it is measured, or it is chosen by drawing from the species–abundance distribution, $\Phi(n)$ which is known from knowledge or estimation of the state variables S_0, N_0. For purposes of testing the prediction for $\Pi(n)$, it is preferable to focus on situations where n_0 is actually measured, but in applications of Eq. 3.13 to predict the shape of the species–area relationship, we often do not have knowledge of all the abundances, and so we rely on draws from the species–abundance distribution.

Referring to the first entry in Table 6.1, and subsequent discussion, the MaxEnt solution to maximizing:

$$I_\Pi = -\sum_{n=0}^{n_0} \Pi(n)\log(\Pi(n)). \tag{7.47}$$

under the constraints of Eqs 7.10 and 7.12, yields the distribution:

Table 7.3 Predicted approximate functional forms for all the metrics derived from $R(n,\epsilon)$. The approximations leading to these entries are valid if $e^{-S_0} \ll 1$. More generally, exact solutions can be derived from the results in Section 7.3.

Metrics derived from $R(n,\epsilon)$	Predicted Function	Properties
Distribution of abundances across species	$\Phi(n\|S_0,N_0) \approx \frac{1}{\log(\frac{1}{\beta})} \cdot \frac{e^{-\beta n}}{n}$	Monotonically declining (logseries)
Distribution of metabolic rates across all individuals in community	$\Psi(\epsilon\|N_0,S_0,E_0) \approx \lambda_2 \cdot \beta \cdot \frac{e^{-\gamma}}{(1-e^{-\gamma})^2}$	Monotonically declining
Distribution of metabolic rates across the individuals in a species with n individuals	$\Theta(\epsilon\|n,S_0,N_0,E_0) \approx \lambda_2 n e^{-\lambda_2 n(\epsilon-1)}$	Boltzmann distribution; "temperature" $\sim (\lambda_2 n)^{-1}$ $= (E_0-N_0)/(S_0 n)$
Dependence on species abundance of metabolic rates averaged over individuals within species	$\bar{\epsilon}(n\|S_0,N_0,E_0) \approx 1 + \frac{1}{n\lambda_2}$	Metabolic rate $\sim n^{-1}$ unless n exceeds $\sim 1/\lambda_2$
Distribution of metabolic rates averaged over individuals within species	$\nu(\bar{\epsilon}\|S_0,N_0,E_0) \approx \frac{1}{\log(\frac{1}{\beta})} \frac{e^{-\frac{\beta}{\lambda_2(\bar{\epsilon}-1)}}}{(\bar{\epsilon}-1)}$	Unimodal, with left–skewed mode (Fig. 7.1); $\beta\lambda_2$ generally $\gg 1$ (see Table 7.2).
Distribution of masses across all individuals in community	$\Xi(m\|S_0,N_0,E_0) \approx$ $\frac{d\epsilon}{dm} \cdot \lambda_2 \cdot \beta \cdot \frac{e^{-\gamma(m)}}{(1-e^{-\gamma(m)})^2}$	Monotonically declining in m if ϵ increases with increasing m; $\gamma(m) = \lambda_1 + \lambda_2 \cdot \epsilon(m)$
Distribution of masses across the individuals in a species with n individuals	$\rho(m\|n,S_0,N_0,E_0) \approx$ $\frac{d\epsilon}{dm} \cdot \lambda_2 \cdot n \cdot e^{-\lambda_2 n(\epsilon(m)-1)}$	Monotonically declining
Dependence of individuals-averaged masses on species' abundance, n	$\bar{m}(n\|S_0,N_0,E_0) \approx \frac{1}{(\lambda_2 \cdot n)^{1/b}} \cdot e^{-\lambda_2 \cdot n}$	Monotonically declining; (Assumes $\epsilon = m^b$)
Distribution of masses averaged over individuals within species	$\mu(\bar{m}\|S_0,N_0,E_0) \approx$ $\frac{1}{\log(\frac{1}{\beta})} \frac{b}{\bar{m}} e^{-\beta/(\lambda_2 \bar{m}^b)}$	Unimodal; (assumes $n\lambda_2 \ll 1$ and $\epsilon = m^b$)

$$\Pi(n) = \frac{1}{Z_\Pi} e^{-\lambda_\Pi n}. \tag{7.48}$$

The normalization summation can be carried out exactly, yielding the partition function:

$$Z_\Pi = \sum_{n=0}^{n_0} e^{-\lambda_\Pi n} = \frac{1 - e^{-\lambda_\Pi(n_0+1)}}{1 - e^{-\lambda_\Pi}}. \tag{7.49}$$

Solving Eq. 7.12 then yields the Lagrange multiplier. To simplify the notation somewhat, we define $x = e^{-\lambda_\Pi}$, leading to:

$$\bar{n} = \frac{n_0 A}{A_0} = \frac{\sum_{n=0}^{n_0} n x^n}{\sum_{n=0}^{n_0} x^n} = \left(\frac{x}{1-x} - \frac{(n_0+1) \cdot x^{n_0+1}}{1 - x^{n_0+1}} \right). \qquad (7.50)$$

Consider the consequences of Eqs 7.48–7.50 for the situation addressed in Section 4.1, where we examined how individuals are allocated between the two halves of a bisected plot: $A = A_0/2$. The solution to Eq. 7.50 is $\lambda_\Pi = 0$ $(x = 1)$, for all n_0, and Z is now equal to $n_0 + 1$. Hence:

$$\Pi(n | \frac{A_0}{2}, n_0, A_0) = \frac{1}{1 + n_0}. \qquad (7.51)$$

In other words, our MaxEnt result is that the probability distribution describing the division of individuals between the two halves of a plot is the uniform distribution. Referring to Section 4.1 and Eq. 4.5, this is precisely the result we obtained from the Laplace Rule of Succession and also from the HEAP model.

Turning to finer spatial scales, we now face a choice. One option is to solve Eq. 7.48 at each scale and thereby determine $\Pi(n)$ at each scale. An alternative is to iterate Eq. 7.49 recursively, just as we did with the HEAP model in Eq. 4.15. In this case, we would obtain the HEAP distribution at all scales. With the first option, we shall see that a particular form of a negative binomial distribution results; this is the truncated negative binomial, which is normalized over a finite range of its argument. The differences between the predictions of HEAP and the truncated negative binomial are very small and generally not empirically distinguishable (see, for example, Figures 4.1a and b). One merit in the non-iterative approach is that by solving Eq. 7.50 at each scale, we allow the option of choosing any arbitrary value of A/A_0; we are not restricting the outcomes to areas, A, that differ from A_0 by powers of 2. This gives us more flexibility and avoids the inconvenience of having to invoke special "user rules" (Ostling et al., 2003) to avoid cross-scale inconsistencies.

The two approaches differ in the way in which prior information is used. The iteration process that results in the HEAP model for $\Pi(n)$, continually "upgrades" information with each iteration. For example, the distribution of abundances, n, at scale $A_0/4$ makes use of the result at scale $A_0/2$, which in turn derived from knowledge only of the total abundance of the species at scale A_0. In contrast, the truncated negative binomial distribution result at scale $A_0/4$, or at any other scale A, derives solely from prior knowledge of the total abundance of the species at scale A_0. This possibility of obtaining different answers to the same question, depending on what prior knowledge is assumed, is a very general feature of MaxEnt calculations;

derived distributions depend upon the nature of the constraints imposed. Does this mean that the HEAP distribution, which iterates the constraint, bisection by bisection and thus in some sense uses more prior information, is more accurate than the truncated negative binomial? Or could it mean that inaccuracies proliferate through the iteration process? We shall return to this issue later.

To explore the form of the truncated negative binomial distribution, consider first the case in which $A \ll A_0$. Numerical solution of Eq. 7.50 reveals that for all values of n_0, $x^{n_0} \ll 1$ and so solutions to that equation are well approximated by:

$$x \approx \frac{\bar{n}}{1+\bar{n}}. \tag{7.52}$$

In the limit in which Eq. 7.52 is valid:

$$\Pi(n) \approx \frac{\left(\frac{\bar{n}}{1+\bar{n}}\right)^n}{1+\bar{n}} = \frac{\left(\frac{n_0 A}{A_0+n_0 A}\right)^n}{\frac{A_0+n_0 A}{A_0}} \tag{7.53}$$

For later use, we note that when $A \ll A_0$:

$$1 - \Pi(0) \approx \frac{n_0}{n_0 + \frac{A_0}{A}} \tag{7.54}$$

to a very good approximation.

The general METE solution for $\Pi(n)$ given by Eq. 7.48, and the particular limiting solution given in Eq. 7.53, are nearly the same as the expression for the negative binomial distribution in Eq. 4.16, for the particular case of Eq. 4.16 in which $k = 1$. The one difference is that in Eq. 7.48, the allowed values of n range from 0 to n_0, while the standard negative binomial distribution is defined over the range from 0 to ∞. The consequences of this distinction between the two binomial distributions has been discussed by Conlisk et al. (2007b).

When A is not $\ll A_0$, and n_0 is sufficiently small, Eq. 7.52 is no longer a good approximation and numerical solutions to Eq. 7.48 are needed. Some worked-out x-values are given in Table 7.4.

We next explore three consequences of our results in Sections 7.3 and 7.4. First we look at the forms of the species–area relationship and the "collector's curve" (Section 3.3.7) that are predicted from the distributions R and Π. Then we examine the implications of the results in Table 7.3 for the energy-equivalence and the Damuth rules that were introduced in Section 3.4.10.

For the same reasons given for testing the probability distributions Φ, Ψ, Θ, and ν using rank-order comparisons, it is preferable to test $\Pi(n)$ against data by comparing the observed and predicted rank versus cell-occupancy values. Box 7.5 derives the predicted rank versus occupancy relationship that is applicable when either $A \ll A_0$, or $n_0 \gg 1$, so that Eq. 7.53 s a good approximation.

Tables 7.4 Values of $x = e^{-\lambda_\Pi}$ calculated by the exact Eq. 7.50 and the approximate Eq. 7.52.

n_0	$A = A_0/4$ Eq. 7.50	$A = A_0/4$ Eq. 7.52	$A = A_0/8$ Eq. 7.50	$A = A_0/8$ Eq. 7.52	$A = A_0/16$ Eq. 7.50	$A = A_0/16$ Eq. 7.52
1	.333	.200	.125	.111	.067	.059
2	.434	.333	.220	.200	.115	.111
4	.568	.500	.344	.333	.201	.200
8	.707	.667	.505	.500	.334	.333
16	.823	.800	.669	.667	.500	.500
32	.901	.889	.801	.800	.667	.667

Box 7.5 The rank–occupancy relationship for $\Pi(n)$

We seek an expression for $n_\Pi(r)$, where $r =$ rank, in analogy to Eqs 7.35 and 7.37. Rank, r, extends from 1 to A_0/A because we are examining the distribution of cell occupancy values over the A_0/A cells of area A. Following the same reasoning as in Section 7.3.1, we write the rank versus cell occupancy relationship as:

$$\sum_{n=0}^{n_\Pi(r)} \Pi(n) = 1 - \frac{(r-0.5)}{A_0/A}. \tag{7.55}$$

Note that the right-hand side of this equation is largest when r is smallest; in other words, we are taking the lowest rank, $r = 1$, to correspond to the largest occupancy value. Assuming $A \ll A_0$ or $n_0 \gg 1$, we can use Eq. 7.53 to derive:

$$\sum_{n=0}^{n_\Pi(r)} \Pi(n) = \frac{1}{1+\bar{n}} \sum_{n=0}^{n_\Pi(r)} \left(\frac{\bar{n}}{1+\bar{n}}\right)^n = 1 - \left(\frac{\bar{n}}{1+\bar{n}}\right)^{n_\Pi(r)+1}, \tag{7.56}$$

where:

$$\bar{n} = n_0 \frac{A}{A_0}. \tag{7.57}$$

Equations 7.54 and 7.56 lead to:

$$n_\Pi(r) = \frac{\log\left(\frac{r-0.5}{A_0/A}\right)}{\log\left(\frac{\bar{n}}{1+\bar{n}}\right)} - 1, \tag{7.58}$$

which can be compared directly with data. Using an Excel spreadsheet to evaluate Eq. 7.58 for various assumed values of n_0 and A_0/A, you can readily show that to ensure that the sum of the predicted abundances is n_0, the values of n_Π obtained from Eq. 7.58 should be rounded up to the nearest integer value. This also ensures that all predicted abundances are non-negative.

When $A < A_0/2$, the quantity $x = \exp(-\lambda_\Pi)$ in Eq. 7.48 is < 1, when $A = A_0/2$, $x = 1$, and when $A > A_0/2$, $x > 1$. Equations 7.50–7.53 are good approximations when $A \ll A_0$, and it turns out there is also a good analytical approximation to $\Pi(n)$ in the other limit in which $A \to A_0$. Because x will greatly exceed 1 in that limit, Eq. 7.50 simplifies to:

$$\bar{n} = n_0 \frac{A}{A_0} \approx \frac{x}{1-x} + n_0 + 1 \qquad (7.59)$$

or

$$x \approx 1 + \frac{1}{n_0(1 - \frac{A}{A_0})}. \qquad (7.60)$$

For $A \sim A_0$, Z is readily shown to be well approximated by $x^{(1+n_0)}/(x-1)$. For later use, we note here that $\Pi(n_0|A, n_0, A_0)$, the probability that all n_0 individuals are found in a cell of area $A = A_0(1 - \delta)$, where $\delta \ll 1$, is:

$$\Pi(n_0|A_0(1-\delta), n_0, A_0) \approx \frac{x-1}{x} \approx \frac{1}{1+\delta \cdot n_0}. \qquad (7.61)$$

Also, $1 - \Pi(0|A, n_0, A_0)$, the probability of presence in A, is readily shown to be

$$\Pi(0|A_0(1-\delta), n_0, A_0) \approx 1 - \frac{1}{x^{n_0}}. \qquad (7.62)$$

7.5 The predicted species–area relationship

Suppose that you wish to know the number of species of plants, or in some other taxonomic group, inhabiting some area, which might be an entire biome, such as the Amazon or the Western Ghats of India, or perhaps just a square kilometer of a particular habitat. Your prior knowledge might consist of the number of species and individuals in a sample of smaller plots within that larger area. This is the problem of up-scaling species' richness. The same theory we use below to up-scale can equally be used to down-scale: to predict the number of species found on average in small plots from knowledge of species' richness in a larger area containing those small plots.

The down-scaling problem is in one sense not very interesting; you probably would not know the number of species in a large area without also knowing the number in small plots nested within that large area. But because we shall use the same theory to both up-scale and down-scale, we can use either up- or down-scaling to test theory. Here we develop the equations that can be used to do either, and then return in Chapter 8 to test the predictions.

Our strategy is to begin with Eq. 3.13, which relates the number of species in area A to the number in area A_0. Whether we are up- or down-scaling, we always use a notation in which $A < A_0$. For down-scaling, we begin with empirical knowledge of S_0 and N_0 in A_0 and then use Eq. 3.13 to predict $\bar{S}(A)$ for $A < A_0$. The notation $\bar{S}(A)$ refers to the average of species' richness over some number of cells of area A. For down-scaling, $\bar{S}(A)$ would be the predicted average over all cells of area A. For up-scaling, $\bar{S}(A)$ would in most practical cases be an average over some subset of cells of area A for which empirical values of species' richness are available and we use Eq. 3.13 to infer $S_0 = S(A_0)$ for values of $A_0 > A$. We call whichever scale at which we have prior empirical knowledge of S and N the "anchor scale."

The two choices we saw in in Section 7.4 for predicting the spatial abundance distribution $\Pi(n)$ now translate into two choices for calculating the form of the SAR. On the one hand (Method 1 below) we can down- or up-scale by doubling or halving area at each step, and re-adjusting the species–abundance distribution using the new information about S and N at each scale change. In that case, we only require knowledge of $\Pi(n|A, n_0, A_0)$ for values of $A = \frac{1}{2} A_0$. An alterative method is to assume the METE form for the species-level spatial abundance distributions $\Pi(n|A, n_0, A_0)$ at every scale ratio, A_0/A, of interest. In this case, we then up- or down-scale using the species-level spatial abundance distributions for arbitrary scale ratios, A/A_0. The details of this second method are presented in Appendix C.

7.5.1 Predicting the SAR: Method I

We first set up the down-scaling problem. Suppose we know there are $S(A)$ species and $N(A)$ individuals in area A and we wish to know the average of the species' richness values in a large number of plots of area $A/2$ nested within A. Note that we are not restricting a plot of area $A/2$ to necessarily be the left or right, or top or bottom, half of A. Because here and in what follows, the Lagrange multiplier, β, will be evaluated at different scales, we write it as $\beta(A)$ to make the scale at which it is evaluated explicit. Note that from Eq. 7.27 or 7.30, $\beta(A)$ is uniquely determined by $S(A)$ and $N(A)$.

If we use Eq. 7.51 to write $\Pi(n|A_0/2, n_0, A_0) = 1 - 1/(n_0+1) = n_0/(n_0+1)$, and substitute that result along with Eq. 7.27 into Eq. 3.13, we can relate $\bar{S}(A/2)$ to $S(A)$ and $N(A)$:

$$\bar{S}(A/2) = \bar{S}(A) \sum_{n=1}^{N(A)} \frac{n}{n+1} \cdot \frac{1}{\log(\frac{1}{\beta(A)})} \cdot \frac{e^{-\beta(A)n}}{n}. \qquad (7.63)$$

Here, for simplicity, we have dropped the subscript on the summation variable n_0. Also note that here we have written $\bar{S}(A)$, not $S(A)$, even though at scale A we assume that there is a specified value for S. The reason is that we will shortly use this expression at arbitrary scale transitions, not just ones in which at the larger scale we have exact information. In those situations, only the average species' richness at both the larger and smaller scales can be known. For convenience we will use the symbol $N(A)$, not $\bar{N}(A)$ at every scale.

Using Eq. 7.27, the summation can be carried out exactly (Exercise 7.8), yielding:

$$\bar{S}(A/2) = \bar{S}(A)e^{\beta(A)} - N(A)\frac{1 - e^{-\beta(A)}}{e^{-\beta(A)} - e^{-\beta(A)(N(A)+1)}}(1 - \frac{e^{-\beta(A)N(A)}}{N(A) + 1}). \quad (7.64)$$

With $\bar{S}(A/2)$ determined, $\bar{S}(A/4)$ then be determined by iteration of Eq. 7.64. To do this requires calculating $\beta(A/2)$, which can be done because it is determined by the now-estimated $\bar{S}(A/2)$ and by $N(A/2)$. N must scale linearly with area, as can be seen by considering that the sum of the abundances in the two halves of A must be $N(A)$. Hence $N(A/2) = N(A)/2$.

When $N \gg \bar{S} \gg 1$, Eq. 7.64 can often be simplified because then $\beta \ll 1$, and $e^{-\beta N} \ll 1$ (see Table 7.1). Hence:

$$\bar{S}(A/2) \approx \bar{S}(A) - N(A)\beta(A). \quad (7.65)$$

Using Eq. 7.30, Eq. 7.65 can be rewritten as:

$$\bar{S}(A/2) \approx \bar{S}(A) - \frac{\bar{S}(A)}{\log(\frac{1}{\beta(A)})}. \quad (7.66)$$

Note that the right-hand side of Eq. 7.66 is always $< \bar{S}(A)$, if $\beta < 1$.

If we now express the species area relationship in the form $\bar{S}(A) \sim A^z$, but allow z to be scale-dependent, we can write:

$$\bar{S}(A) = 2^{z(A)}\bar{S}(A/2). \quad (7.67)$$

Then from Eq. 7.66, it follows (Exercise 7.9) that:

$$z(A) \approx \frac{\log(\frac{\log(1/\beta(A))}{\log(1/\beta(A))-1})}{\log(2)}. \quad (7.68)$$

If β is sufficiently small, so that $\log(1/\beta) \gg 1$, then Eq. 7.68 can be further simplified:

$$z(A) \approx \frac{1}{\log(2) \cdot \log(1/\beta(A))}. \quad (7.69)$$

Provided $S \gg 1$ and $\beta \ll 1$ remain true as Eq. 7.64 is iterated to finer scales, Eq. 7.69 remains valid and it provides a general expression for the scale dependence of z. Because $\beta(A)$ is, to a very good approximation, a function only of the ratio $\bar{S}(A)/N(A)$, if the inequalities hold (see Table 7.2), Eq. 7.69 informs us that species–area relationships should all collapse on to a universal curve if $\log(\bar{S}(A))$ is plotted against the variable $\log[N(A)/\bar{S}(A)]$ instead of against the usual variable $\log(A)$. Stated differently, METE predicts that the value of $N/\bar{S}$ at some specified scale determines the shape of the SAR at larger or smaller scales.

Figure 7.2 illustrates this concept of scale collapse. The large graph shows three made-up species–area curves, all plotted as $\log(S)$ versus $\log(\text{area})$. They clearly have different shapes, and in fact at any given value of area, the largest and smallest slopes differ by almost a factor of two. The curves differ by amounts that are not atypical of actual SARs. In parentheses, the values of S and N are given for one data point on each curve. These three data points were chosen because they all share a common value of N/S (equal to 50). The inset graph plots slope z against $\log(N/S)$. The filled circle corresponds to all three of the data points with $N/S = 50$, the ones labeled on the larger graph. The three data points "collapse" onto a single data point, the filled circle, when plotted against $\log(N/S)$ rather than against area. The unfilled circles on the inset graph correspond to other groups of points on the SARs with common values of N/S.

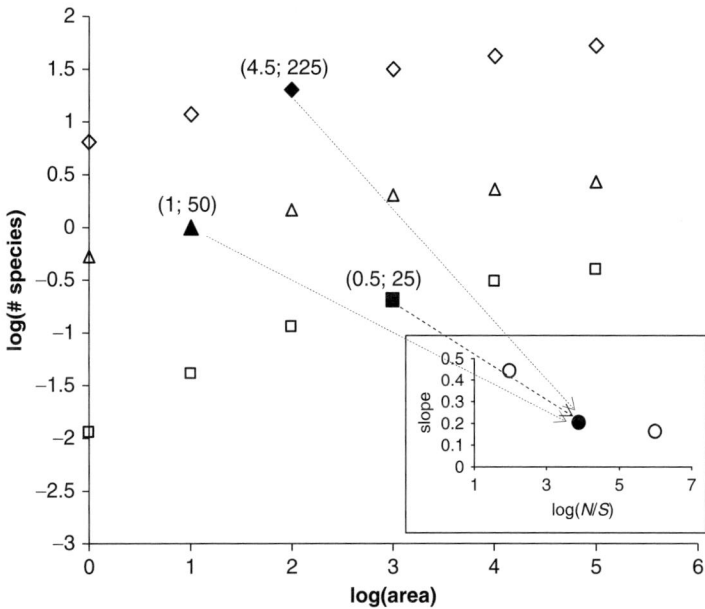

Figure 7.2 The concept of scale collapse. The main graph shows three SARs, each with a different shape symbol. The SARs appear to be unrelated to each other. The labels (x, y) on the filled data point on each SAR give the values of S and N for that.datum. The data points that were filled were selected because the slope of each SAR (tangent to the curve) is the same (0.21) for all three curves at those points. The inset graph shows slope, z, versus $\log(N/S)$ at three values of N/S. The filled data point on the inset corresponds to the three filled data points on the SARs. METE predicts that all SARs share a common slope at a common value of N/S, and that is the meaning of scale collapse.

The essential point is that all SARs are predicted to have the same shape, differing only in the scale at which they share a common $N/\bar{S}$ value and the actual number of species at that value. Thus we predict that all SARs are scale displacements of a universal SAR shape. Chapter 8 tests this prediction.

The scaling relationship for $z(A)$ is of particular interest when Eq. 7.64 is used for up-scaling, because then the requisite inequalities hold to a better and better approximation as area increases. The reason is that S is now increasing with each iteration, and because N increases linearly with area, while $\bar{S}$ increases less steeply, the ratio $\bar{S}/N$, and hence β, decreases with increasing area. So, if the inequalities are satisfied at some anchor scale, they will hold increasing well as area increases.

The up-scaling procedure is slightly more complicated than the down-scaling procedure because now both Eq. 7.27 and Eq. 7.64 have to be solved simultaneously. To see why, assume that from censusing a sample of plots of scale A within some much larger biome of area A_0, we have an estimate of $\bar{S}(A)$ and $\bar{N}(A)$. We now wish to estimate $\bar{S}(2A)$, so we iterate Eq. 7.64 to give us:

$$\bar{S}(A) = \bar{S}(2A)e^{\beta(2A)} - N(2A)\frac{1 - e^{-\beta(2A)}}{e^{-\beta(2A)} - e^{-\beta(2A)(N(2A)+1)}}\left(1 - \frac{e^{-\beta(2A)N(2A)}}{N(2A) + 1}\right). \quad (7.70)$$

To determine $\beta(2A)$, we use Eq. 7.27:

$$\frac{\bar{S}(2A)}{N(2A)} \sum_{n=1}^{N(2A)} e^{-\beta(2A)n} = \sum_{n=1}^{N(2A)} \frac{e^{-\beta(2A)n}}{n}. \quad (7.71)$$

Equations 7.70 and 7.71 comprise two coupled equations for the two unknowns $\beta(2A)$ and $\bar{S}(2A)$ and can be readily solved numerically. The knowns are $\bar{S}(A)$ and $N(2A) = 2N(A)$. Iteration can be carried out up to an arbitrarily large spatial scale. Excel can no longer do the job, but the other mathematical software listed above can.

Equations 7.70 and 7.71 are the equations we will use to test the SAR prediction from METE. However, an alternative procedure is possible (method 2) based on using the species-level spatial abundance distribution at arbitary scale, rather than just assuming its results for a single bisection. The lengthy details are presented in Appendix C. It is worth noting that the results for method 2 differ only slightly from the method 1 results.

7.5.2 The special case of $\bar{S}(A)$ for $1 - A/A_0 \ll 1$

For areas, A, that are only slightly smaller than A_0, we can approximate the dependence of species' richness on area. This, and the related endemics–area relationship (Section 7.7), are of interest for evaluating species' richness in areas that remain after habitat is lost.

How does species' richness depend on area for areas, A, that are only slightly smaller than A_0, or in other words for the case in which $A = A_0(1 - \delta)$, where $\delta \ll 1$? From Eqs 3.13, 7.68, and 7.70, the number of species in a cell with area A, can be approximated as:

$$\bar{S}(A) \approx \frac{\bar{S}(A_0)}{\log(\frac{1}{\beta})} \sum_{n=0}^{N_0} \frac{e^{-\beta n}}{n} \left(1 - \frac{1}{(1 + \frac{1}{\delta \cdot n})^n}\right). \tag{7.72}$$

This complicated-appearing summation is intractable analytically, but numerical evaluation of the result for a variety of values of δ, β, and N_0 reveals that, to a very good approximation, the following result holds:

$$S(A) \approx \bar{S}(A_0)\left(1 - \frac{\delta}{\log(\frac{1}{\beta})}\right). \tag{7.73}$$

Thus, for $\delta = 1 - A/A_0 \ll 1$, $\bar{S}(A)$ is a linear function of A/A_0 with proportionality constant equal to $1/\log(1/\beta)$. The A-dependence of $\log(1/\beta)$ is relatively weak.

7.6 The endemics–area relationship

The endemics–area relationship, $\bar{E}(A)$, was defined in Section 3.3.8. Equation 3.8 determines the relationship in terms of the species-level spatial abundance distributions, $\Pi(n)$, and the species–abundance distribution, $\Phi(n)$. For $A \ll A_0$, so that Eq. 7.51 is valid, we can substitute Eqs 7.30 and 7.51 into Eq. 3.8, to obtain:

$$\bar{E}(A) \approx \frac{S_0}{\log(\frac{1}{\beta})} \sum_{n=1}^{N_0} \frac{e^{-\beta n}}{n} \frac{A_0}{nA + A_0} \left(\frac{nA}{nA + A_0}\right)^n. \tag{7.74}$$

Where the $\approx$ sign is used because we are assuming that $\beta N \gg 1$ and $\exp(-S_0) \ll 1$ (so that the normalization constant in the logseries distribution is well approximated by $1/\log(1/\beta)$, and also that $A \ll A_0$.

A good analytical approximation to Eq. 7.74, valid if $A \ll A_0$, is:

$$\bar{E}(A) \approx \frac{AS_0}{A_0 \log(\frac{1}{\beta})}. \tag{7.75}$$

In practice, Eq. 7.75 is a quite accurate approximation to Eq. 7.74 if $A < A_0/10$.

Another limit in which we can find a useful analytical approximation to the EAR is for values of A that exceed $A_0/2$. Letting $A = A_0(1 - \delta)$, where $\delta \ll 1$, and using Eq. 7.59, we have:

$$\bar{E}(A) \approx \frac{S_0}{\log(\frac{1}{\beta})} \sum_{n=1}^{N_0} \frac{e^{-\beta n}}{n} \frac{1}{1+\delta \cdot n}. \tag{7.76}$$

Although this summation cannot be carried out in closed form, if $\beta < \delta$, then a good approximation to Eq. 7.74 is:

$$\bar{E}(A) \approx \frac{S_0}{\log(\frac{1}{\beta})} \log(\frac{1}{\delta}). \tag{7.77}$$

Because for actual datasets, β is nearly always < 0.01, this approximation is usually valid for areas A that are as large as 99% of the total area A_0.

Finally, we observe that another way to derive the EAR is to note that if A_0 is divided into two pieces, of area A and $A_0 - A$, then $S(A0) = \bar{E}(A) + \bar{S}(A_0 - A)$. Hence, $\bar{E}(A) = S(A_0) - \bar{S}(A_0 - A)$. Using our derived expression for the SAR, the EAR is determined. A problem with this approach arises when method 1 (Section 7.5.1) is used to calculate the SAR. If the cell of area A is a small plot in the center of a large plot of area A_0, then the area $A_0 - A$ is doughnut-shaped. The theory for scaling species' richness based on doubling or halving the size of some initial plot will never produce such shapes. But for the situation in which $A = \frac{1}{2} A_0$, and the two A-cells are taken to be the two halves of A_0, then this approach works, and we obtain:

$$\bar{E}(A_0/2) = S(A_0) - \bar{S}(A_0/2). \tag{7.78}$$

If $S(A_0/2)$ and $\bar{E}(A_0/2)$ and are evaluated using Eqs 3.13 and 3.14, with the exact expression for $\Pi(n|A_0/2, n_0, A)$, it can be shown that Eq. 7.78 is an identity (Exercise 7.10).

7.7 The predicted collector's curve

We can also extract the shape of the collector's curve from the predicted form of the species' abundance, Eq. 7.32. Recall that the collector's curve describes the rate of accumulation of species as individuals are randomly sampled from the individuals' pool. If sampling is carried out without replacement, then the shape of the collector's curve, $\bar{S}(N)$ is identical to the shape of the species–area relationship, $\bar{S}(A)$, that would be obtained if individuals were randomly distributed on a landscape. The reason is that sampling a larger area is the same as sampling a larger number of individuals.

In Box 7.6, we show that for the case $N/N_0 \ll \beta_0$, corresponding to the sampling of only a small fraction of all the individuals, species accumulate at a rate that is equal to sample size: $\bar{S}(N) \approx N$ (Eq. 7.86).

At larger N, with $N/N_0 \gg \beta_0$, but N/N_0 still $\ll 1$, we derive:

$$\bar{S}(N) \approx a_1 + a_2 \log(N), \tag{7.79}$$

where

$$\begin{aligned} a_1 &= S_0\left(1 - \frac{\log(N_0)}{\log(1/\beta_0)}\right), \\ a_2 &= \frac{S_0}{\log(1/\beta_0)}. \end{aligned} \tag{7.80}$$

Using Eq. 7.30, we see that Eq. 7.80 can be rewritten as $a_2 = \beta_0 N_0$, showing that the dependence of the collector's curve on N at intermediate values of N and the dependence of species' richness on area are, respectively, proportional to $\log(N)$ and $\log(A)$.

To summarize the predicted collector's curve: as N increases, the collector's curve transitions from linear to logarithmic.

Finally, as $N \to N_0$, the full expression in Eq. 7.84 in Box 7.6 must be used.

7.8 When should energy-equivalence and the Damuth relationship hold?

The notion of energy-equivalence was introduced in Section 3.4.10. One formulation of energy-equivalence asserts that the average metabolic rate of the individuals in a species with abundance n is proportional to $1/n$, implying that the total metabolic rate of all individuals in a species is independent of its abundance. In other words, under energy-equivalence all species utilize equal shares of the total E_0. The related Damuth rule, asserting that the mass of individuals in a species varies as the inverse 4/3 power of the abundance of the species, would follow from the formulation of energy-equivalence given above, if the metabolic scaling rule $\epsilon \sim m^{3/4}$ were valid.

Empirical support for either energy-equivalence or the Damuth rule is mixed. Patterns in the relationship between organism mass and abundance were reviewed by White et al. (2007). Their findings were summarized in Section 3.4.10, but interested readers should refer to their paper for details. Energy-equivalence has also been examined for plant communities by Enquist and colleagues (Enquist et al., 1999; Enquist and Niklas, 2001). They estimate metabolic rate from organism mass using the ¾ power metabolic scaling rule. To estimate the mass of, say, a tree, they assume an allometric relationship between a quantity such as basal area, which is relatively easy to measure, and tree volume, which is not. Thus the validity of energy-equivalence becomes entangled with the validity of both the metabolic scaling law and the assumed allometric relationship. Ubiquitous and consistent support for either energy-equivalence or the Damuth rule is lacking, and yet there does appear to be support for the notion that there is an inverse relationship between abundance and either mass or energy, for at least many, if not all, groups of taxa, types of ecosystems, and scales of analysis.

Box 7.6 Derivation of the collector's curve

We start with the formula (Eq. 3.13) for the SAR in terms of the spatial distribution Π, and $\Phi(n)$, the species–abundance distribution. Then, using the Coleman random placement result for $\Pi(0)$ (Eqs 4.1 and 4.2), and the METE prediction for $\Phi(n)$ (Eq. 7.32), we get for the shape of the collector's curve:

$$\bar{S}(N) = S_0 \sum_{n=1}^{N_0} [1-(1-\frac{N}{N_0})^n] \frac{e^{-\beta_0 n}}{n \cdot \log(1/\beta_0)}. \tag{7.81}$$

Using $a^b = e^{b \cdot \log(a)}$, and the normalization condition on $\Phi(n)$, this can be rewritten in the form:

$$\bar{S}(N) = S_0 \sum_{n=1}^{N_0} [1 - e^{n \cdot \log(1-N/N_0)}] \frac{e^{-\beta_0 n}}{n \cdot \log(1/\beta_0)} = S_0 - S_0 \sum_{n=1}^{N_0} \frac{e^{-n[\beta_0 - \log(1-N/N_0)]}}{n \cdot \log(1/\beta_0)}. \tag{7.82}$$

Because $N_0 \beta_0 \gg 1$, it follows that $N_0[\beta_0 - \log(1-N/N_0)] \gg 1$, and so the summation on the right-hand side of Eq. 7.82 can be carried out to a good approximation:

$$\bar{S}(N) \approx S_0 - S_0 \frac{\log(1/[\beta_0 - \log(1-N/N_0)])}{\log(1/\beta_0)}. \tag{7.83}$$

Using $\log(a \cdot b) = \log(a) + \log(b)$, and $\log(a) = -\log(1/a)$, this becomes:

$$\bar{S}(N) \approx S_0 + S_0 \frac{\log[\beta_0 - \log(1-N/N_0)]}{\log(1/\beta_0)} = S_0 + S_0 \frac{\log[\beta_0(1-(1/\beta_0)\log(1-N/N_0)]}{\log(1/\beta_0)}$$
$$= S_0 \frac{\log[1-(1/\beta_0)\log(1-N/N_0)]}{\log(1/\beta_0)}. \tag{7.84}$$

Using $\log(1+a) \approx a$ if $a \ll 1$, we can simplify Eq. 7.84 if $N/N_0 \ll 1$:

$$\bar{S}(N) \approx S_0 \frac{\log[1 + N/(\beta_0 \cdot N_0)]}{\log(1/\beta_0)}. \tag{7.85}$$

We now consider two cases. First let $N/N_0 \ll \beta_0$. Then:

$$\bar{S}(N) \approx \frac{S_0 \cdot N}{N_0 \cdot \beta_0 \cdot \log(1/\beta_0)} \approx N, \tag{7.86}$$

where we have used Eq. 7.30 to simplify the expression above. Because β_0 is generally much smaller than 1, this limit corresponds to very small values of N/N_0, and thus applies to the first individuals collected. Thus the collector's curve starts out linear in N; each new individual is, at first, likely to be selected from a different species.

Second, at larger N, with $N/N_0 \gg \beta_0$, but N/N_0 still $\ll 1$, we have from Eq. 7.84:

Box 7.6 (*Cont.*)

$$\bar{S}(N) \approx S_0 \frac{\log[[N/(\beta_0 \cdot N_0)]]}{\log(1/\beta_0)} = a_1 + a_2 \log(N), \quad (7.87)$$

where:

$$a_1 = S_0\left(1 - \frac{\log(N_0)}{\log(1/\beta_0)}\right),$$
$$a_2 = \frac{S_0}{\log(1/\beta_0)} \quad (7.88)$$

Hence, as N increases, the collector's curve transitions from linear to logarithmic. Finally, as $N \to N_0$ the full expression in Eq. 7.84 must be used.

METE predicts the conditions under which energy-equivalence should or should not be valid. To see this, start by multiplying the expression for $\Theta(\epsilon|n)$ in Eq. 7.34 (the distribution of metabolic rates across individuals of a species with abundance n) by $n*\epsilon$ and integrating over ϵ, to derive an expression for the total metabolic rate for all the individuals in a species with abundance n:

$$n\bar{\epsilon} \approx n + \frac{1}{\lambda_2}. \quad (7.89)$$

Eq. 7.89 bears a resemblance to Eq. 7.26, rewritten as $E_0/S_0 = N_0/S_0 + 1/\lambda_2$. This is because E_0/S_0 is the average total metabolic rate per species, and N_0/S_0 is the average abundance per species. The two expressions assert different things, however; Eq. 7.26 is a community-level relationship that determines the value of the Lagrange multiplier λ_2 from the state variables S_0 and E_0, while Eq. 7.89 tells us that if a species with abundance n is selected from the species list, and if $n \ll 1/\lambda_2$, then the total metabolic rate of all the individuals in that species, $n\bar{\epsilon}$, is independent of n. In other words, the subset of all the species that have abundances $n \ll 1/\lambda_2$ all have the same total metabolic rate, $1/\lambda_2$. This is the energy-equivalence principle discussed in Section 3.4.10.

The necessary and sufficient condition for energy-equivalence to hold, that $n\bar{\epsilon}$ is independent of n, can be re-expressed as $1/\lambda_2 = (E_0 - N_0)/S_0 \gg n$. So we can now formulate the prediction:

1. The class of species that obey energy-equivalence is the class of species with abundance $n \ll 1/\lambda_2 = (E_0 - N_0)/S_0$.

We can go further, however. In general, for any ecosystem we would not expect there to be any necessary relationship between the distribution of metabolic rates across the species from a taxonomic group found in the ecosystem, and across the individuals in those species, but in fact METE does lead to such a relationship. To see this, again consider the function $\Theta(\epsilon|n)$. Equation 7.34 informs us that it is an exponentially decreasing function of metabolic rate, ϵ, with the decay constant given by $\lambda_2 n$. Thus, the bigger $\lambda_2 n$, the steeper is the distribution, and when $\lambda_2 n \ll 1$, the distribution will be relatively flat. In other words, the metabolic rates of the individuals in the species with abundance $n \ll 1/\lambda_2 = (E_0 - N_0)/S_0$ will span a wide range of values. These are the species predicted to obey energy-equivalence. In contrast, for species in which $n \gg 1/\lambda_2 = (E_0 - N_0)/S_0$, and energy-equivalence fails, the metabolic rates of the individuals will be clustered around small values.

The latter case might seem like it is trivially forced by energy conservation. To see that it is not, consider the following hypothetical set of state variables: $S_0 = 200$, $N_0 = 10{,}000$, $E_0 = 20{,}000$. For that community, $\lambda_2 = 200/(20{,}000 - 10{,}000) = 0.02$. Within the community, a species with abundance $n = 2000$ could exist, and for that species $\lambda_2 n = 40 \gg 1$. Hence the metabolic rates of the individuals are predicted to all cluster tightly around the minimum value $\epsilon = 1$. An individual with $\epsilon = 2$ would have a probability of less than e^{-40} of the probability of an individual with $\epsilon = 1$. Yet, in principle, were METE to fail significantly, the species could have individuals with metabolic rate as high as $\sim E_0 - N_0 \epsilon_{min} = 10{,}000$.

So we are led to the second prediction:

2. The class of species that should obey energy-equivalence is the class of species whose individuals span a relatively wide range of metabolic rates.

Body sizes and metabolic rates of individuals are strongly and positively correlated, at least approximately by a power-law scaling relationship $\epsilon = m^b$, with evidence for the special value $b \approx 3/4$. Assuming only a general positive relationship between mass and metabolism, our two predictions can be re-expressed with body size substituted for metabolic rate. Stated that way, we can then make a third prediction, which follows from the first two, and surprisingly relates intra-specific variability in body size to interspecific variability:

3. Communities in which there is large (small) size variation across species will be comprised of species with large (small) size variation across life stages.

While these three predictions are formulated somewhat qualitatively, with phrases like "differ most," "wide range," the explicit forms of the mass and energy metrics in Table 7.3 provide the precision that can in principle be used for hypothesis testing. Given the complexity of ecosystems, the imprecision embodied in most macroecological datasets, and the unlikelihood that our MaxEnt theory could possibly be accurate to high precision, I suspect that the qualitative formulations are all that can reasonably be expected to be validated.

Note that even if $E_0 \gg N_0 \gg S_0$, however, so that λ_2 is small compared to $S_0/N_0 = 1/<n>$ there could be some species in the system with abundance n that is comparable to or greater than $1/\lambda_2$ and for those species energy-equivalence is predicted not to hold.

The bottom line here is that the energy-equivalence (or the Damuth rule) should not hold for species in which metabolic energy rates (or masses) vary only a little across individuals within a species, and that class of species should exhibit relatively low variability in body size across species. Those species should be outliers on graphs showing the central tendency in the relationship between mass and abundance or metabolic rate and abundance.

From an evolutionary/physiological perspective this predicted connection between taxa that vary in size over their lifespan, and taxonomic groups within which species vary greatly in size, makes sense. Species whose individuals vary greatly in size over the life cycle of an individual (trees are the best example) clearly possess the capacity to function over a wide range of body sizes under the set of constraints and advantages that result from being that species. From this we may infer that if a lineage of such a species splits, individuals in the resulting two lineages could, in principle, differ somewhat more in body size (at a comparable level of ontogeny) than would individuals in the two lineages resulting from a species with a tightly size-constrained body plan. Thus species radiating from the former type of species would exhibit a wider range of average body sizes.

7.9 Miscellaneous predictions

In Sections 7.4–7.9 the major predictions of METE are presented: the species-level spatial abundance distribution, $\Pi(n)$, across spatial scales; the species–area relationship, $S(A)$; the species–abundance distribution, $\Phi(n)$; the collector's curve; and all the energy and mass metrics listed in Tables 3.2 and 3.3. Not discussed were:

- Range–abundance distributions, $B(A|n_0, A_0)$.
- Metrics that describe correlations, including $C(A, D)$, $\bar{X}(A, D|S_0, A_0)$, Ripley's K, and the O-ring metric.
- Linkage distributions, $\Lambda(l)$.
- Dispersal distributions, $\Delta(D|n_0, A_0)$.

Predictions for the first of these are readily derived from $\Pi(n)$ and $\Phi(n)$. The form of $B(A)$ follows immediately from Eq. 3.7.

I return to the question of what MaxEnt can inform us about spatial correlations in Chapter 11.

Linkage distributions in any sort of network, including trophic webs or plant–pollinator networks, can be inferred from MaxEnt if prior knowledge of the total number of nodes (S_0, for example, in a food web) and the total number of linkages, L_0, are known. Then, from the mean number of linkages per node, exponential linkages distributions are predicted (Williams, 2010). That is, the distribution $\Lambda(l) \sim e^{-\lambda_\lambda l}$ is predicted.

In some cases more detailed knowledge of the number of species at different trophic levels might be available, and then, if the total number of linkages connecting to each trophic level is also known, distinct exponential distributions for the numbers of linkages at each level are predicted.

If dispersal distributions are assumed to determine, mechanistically, the observed or predicted shapes of species-level spatial abundance distributions, $\Pi(n|A, n_0, A_0)$, then the predicted shapes of dispersal distributions can be inferred. An example of this, for the HEAP model spatial abundance distributions, is given in Harte (2006). Hence, for the iterated METE prediction for $\Pi(n)$, which is just the HEAP model prediction, dispersal distributions are predicted. A similar prediction for the non-iterated (bounded negative binomial) $\Pi(n)$ could, in principle, be derived.

7.10 Summary of predictions

Here is a summary of the major predictions of METE.

1. The species-level spatial abundance distribution is identical to the uniform distribution predicted by the HEAP at the spatial scale of a cell bisection: $\Pi(n|A = A_0/2, n_0, A_0) = (1+n_0)^{-1}$. At finer scales, METE predictes either an exponential (Boltzman) distribution or, if the bisection result is iterated, the HEAP distribution. From the abundance distribution the collector's curve is derived: $S(N) = N$ for $N/N_0 \ll \beta$ and $S(N) \sim \log(N)$ for $\beta < N/N_0 \ll 1$.
2. Species–abundance distribution. $\Phi(n)$ is predicted to be a Fisher logseries distribution: $\Phi(n|S_0, N_0) = (c/n) \cdot \exp(-\beta \cdot n)$. Here c is a normalization constant and β is a function of S_0 and N_0 given by the solution to the exact Eq. 7.27, and usually to a good approximation by the solution to Eq. 7.30. The number of species predicted to have a single individual is $\beta \cdot N_0$, while the number predicted to have less than or equal to n_c individuals (when βn_c is $\ll 1$) is βN_0 times the sum of the inverses of the first n_c integers, or approximately $\beta \cdot N_0 \cdot (\log(n_c) + 1)$.
3. Species–area relationship. If the SAR is written in the form $S(A) = c \cdot S^{z(A)}$, $z(A)$ is predicted to depend on the state variables at scale A as shown in Eq. 7.68. Recalling that $N(A)$ is propotional to A, this gives rise to a universal shape for the SAR, in which plots of $S(A)$ against N/S exhibit a scale collapse onto a universal curve. This notion of scale collapse is explained in Figure 7.2.
4. Metabolic rate distributions. The predicted rank–metabolism relationship for all the individuals in a community is given by Eq. 7.37. The distribution of metabolic rates over the individuals within each species is a Boltzman distribution (Eq. 7.34), which decreases monotonically with metabolic rate; in contrast, the distribution across species of the intra-specifically-averaged metabolic rates is unimodal (Eq. 7.46). The most energetic individual in the

community is predicted to have a metabolic rate given by $\epsilon_{\Psi,\max} = \log(2\beta N_0)/\lambda_2$. Predicted mass distributions will depend on the form of the relationship between mass and metabolic rate, $\epsilon(m)$.

5. Energy-equivalence principle. The product of abundance times average metabolic rate is predicted to be independent of abundance for those species with abundance $n_0 < (E_0 - N_0)/S_0$. Communities of plants or mammals or birds in which there is large (small) size variation across species will be comprised of species with large (small) size variation across life stages.

7.11 Exercises

Exercise 7.1

Derive Eqs 7.16 and 7.17.

Exercise 7.2

Derive Eq. 7.19.

Exercise 7.3

Derive Eq. 7.20.

Exercise 7.4

Show that Eq. 7.26 follows from Eqs 7.19 and 7.20, if all terms with $e^{-\sigma}$ are dropped.

Exercise 7.5

Explain why λ_2 as given by Eq. 7.26, is never negative (in other words explain why N_0 can never exceed E_0).

Exercise 7.6

Derive Eqs 7.32–7.34 under the assumptions stated in the text.

Exercise 7.7
Derive Eqs 7.36–7.38.

Exercise 7.8
Derive Eq. 7.64.

Exercise 7.9
Derive Eq. 7.68.

Exercise 7.10
Show that Eq. 7.78 is an identity.

8

Testing METE

Numerous tests of METE predictions for spatial abundance distributions, community-level abundance distributions, species–area relationships, and endemics–area relationships were presented in Harte et al. (2008, 2009). Only a few of these are included here; the emphasis is on predictions of METE that were not hitherto tested. While observed patterns in macroecology generally resemble the predictions of METE, deviations do arise; systematic trends in these deviations are noted and their implications are discussed.

8.1 A general perspective on theory evaluation

Evaluating the absolute or the relative-to-other-theories "goodness" of a theory is not an easy task. Theories differ, in the number of adjustable parameters that they contain, and thus in the degree to which they can be flexibly tuned to fit empirical data. They also differ in their comprehensiveness... the number of distinct phenomena that they purport to describe. For those reasons, comparisons among different theories, leading to determination of the "best" one, are not straightforward. Statistical procedures do exist for taking into account the number of fitting parameters, with the aim of penalizing highly parameterized models or theories (see, for example, Akaike, 1974), but there is no single, unambiguously best procedure for doing so. Moreover, no formal procedures currently exist for evaluating the best theory among a group of theories that predict the forms of differing numbers of metrics with differing levels of accuracy. Thus, there is no agreed-upon method for deciding whether a theory that predicts the form of, say, two metrics very accurately or one that predicts the form of five metrics a little less accurately, is a better theory.

Finally, theory evaluation in ecology is rendered difficult because no criteria exists for "how good a fit is good enough," at least not one that all ecologists can agree upon. One reason for this is that most datasets in macroecology come with imprecisely understood sources of uncertainty and thus ill-characterized magnitudes of error. This is due to: (a) the difficulty of truly replicating an ecological phenomena; (b) errors inevitably present in census data obtained by fallible observers; (c) stochasticity in the influences of the physical environment on populations; (d) stochasticity, even under a constant environment, in the demographics of those populations. A second reason for the absence of such criteria is that, in conservation biology and in other applied fields where macroecological predictions are needed, we lack a quantitative sense of how accurate our predictions have to be to warrant basing policy responses upon them. This contrasts with, for example, toxicology, where regulatory agencies often have agreed-upon criteria for how to respond to results of laboratory testing.

For all the above reasons, I am not going to try to convince you that METE predicts SARs or SADs or energy distributions better, i.e. more accurately, than do other currently deployed theories and models. In fact, there are certainly fitting functions that will do a better job than METE describing any particular one of the metrics that METE predicts. My aim here is to examine the degree to which METE does capture the central tendencies, if not all the details, of the empirical patterns of macroecology.

It is important to understand that METE contains no adjustable fitting parameters. In other words, once the state variables, S, N, E are measured at some spatial scale A, then all the metrics listed in Table 3.1 (cell–occupancy distributions, species–area and endemics–area relationships, species–abundance distributions, distributions of metabolic rates over individuals and species) are predicted at multiple spatial scales. Moreover, the mass-metrics in Table 7.3 are then also uniquely predicted if the dependence of metabolism on body mass is specified.

Two difficult questions should be kept in mind as we compare the predictions of METE with data: are the patterns in nature predicted accurately enough to be of practical use (say in extrapolating species' richness from small to large spatial scales) and are there systematic trends in the deviations between observation and prediction that may provide clues to either future refinements of the theory or to the possibility that some explicit mechanism, not captured by MaxEnt, is needed to explain those deviations. In that last case, the analogy with corrections to the ideal gas law that was discussed in Chapter 1 is apt.

8.2 Datasets

To test our predictions, we have examined census data for plants, birds, arthropods, mollusks, and microorganisms. The data come from habitats that include wet and dry tropical and subtropical forests, subalpine meadows, serpentine grasslands, desert, temperate forest understory, and savannah. With a few exceptions, the datasets are from relatively undisturbed sites; future tests of theory from highly disturbed sites, such as in the aftermath of wildfire or avalanche, where perhaps rapid ecological recovery toward a pre-disturbed system might be occurring, would be highly desirable. The sites from which we have data suitable for testing METE, along with some basic characteristics of those sites, are listed in Table 8.1.

Two measures of the spatial scale of the datasets are useful, even though precise definitions of these measures are often not possible. The first is the size of a sample unit, such as a quadrat or a soil core, in which taxa are censused (a spatial scale called "grain size"). This is a spatial scale that defines the limit of resolution of locational data. The second is the area over which the "grain" are located (a spatial scale sometimes called "extent"). The data we examine here are taken from census

Table 8.1 Sites from which census data are used for testing METE.

Site & taxa	Area(ha)	S_0	N_0	β
Anza Borrego: desert plants[a]	0.0016	24	2445	0.00146
BCI-2000: tropical trees[b]	50	302	213791	0.000162
BCI-2005: tropical trees[b]	50	283	208310	0.000155
Bukit Timah: tropical trees[b]	2	296	11843	0.00464
Claiborne Bluff Mollusks[c]	n.a.	51	1448	0.007085
Cocoli: tropical trees[b]	2	130	4807	0.00511
Desert Grasshoppers[d]	n.a.	22	1378	0.00261
Gothic Earthflow: subalpine flora[e]	0.0064	27	1626	0.00279
HI arthropods: KA[f]	n. a.	158	1922	0.021
HI arthropods: KH[f]	n. a.	240	6048	0.00823
HI arthropods: LA[f]	n. a.	156	2253	0.0168
HI arthropods: MO[f]	n. a.	227	3865	0.0135
HI arthropods: VO[f]	n. a.	167	1909	0.0228
Korup: tropical trees[b]	50	495	329026	0.000174
La Planada: tropical trees[b]	25	221	105163	0.000254
Lambir: tropical trees[b]	52	1161	355327	0.00042
Luquillo: tropical trees[b]	16	137	67465	0.000244
Mudumalai: tropical trees[b]	50	67	17995	0.000488
Pasoh: tropical trees[b]	50	808	296124	0.000341
San Emilio: dry tropical trees[g]	9.8	138	12851	0.00168
Serpentine 1998: meadow flora[h]	0.0064	24	37182	5.92E-05
Serpentine 2005: meadow flora[i]	0.0064	28	60346	4.27E-05
Sinharaja: tropical trees[b]	25	205	193353	0.000117
South African birds[j]	n.a.	47	264	0.061
Subalpine forest understory flora[k]	0.0001	12	547	0.003175
Temperate invertebrates[l]	0.0064	11	414	0.003872
Western Ghats trees (I)[m]	0.25	32.5	109	0.134
Western Ghats trees (II)[n]	1	49	645	0.0189
Yasuni: tropical trees[b]	50	1084	145575	0.00109

Data ownership and sources:

[a] Data collected by J. and M. Harte (unpublished).
[b] Smithsonian Tropical Research Institute data: Condit (1998); Hubbell et al. (1999, 2005).
[c] Data from Harnik et al. (2009) and unpublished.
[d] Data from Rominger et al. (2009).
[e] Data from E. Newman, J. Harte, A. Messerman, and D. Bartholemew, unpublished.
[f] Data from Gruner (2007) and unpublished.
[g] Data from Enquist et al. (1999).
[h] Data from Green et al. (2003)
[i] Data from A. Smith and D. Christiansen, unpublished.
[j] Data from D. Storch, described in Harte et al. (2009).
[k] Data from Harte et al. (2009).
[l] Data from J. Goddard, unpublished.
[m] Data from Krishnamani et al. (2004).
[n] Data from Ramesh et al. (2010).

designs with grain sizes ranging from fractions of a gram of soil and the nearly point locations of individual plant stems, up to plots of area 1 ha. The extents vary over a huge scale range from 1 m^2 to ¼ ha to 50 ha to 60,000 km^2. In some locations, there are really three spatial scales characterizing a site. Consider, for example, the two datasets, Western Ghats (I) and (II), in Table 8.1. In Western Ghats (I), trees locations are resolved to a grain size of 20 m^2, within each of 48 plots of area ¼ ha, and the plots are distributed over 60,000 km^2. In Western Ghats (II), tree locations are resolved to 1 ha and 96 such 1-ha plots are distributed over 28,000 km^2.

8.2.1 Some warnings regarding censusing procedures

Censusing total abundance of plants poses certain unique problems. The main problem, resolving what is an individual, is generally settled by the convention of counting genets, although ramets might be counted for clonal species. For tree censuses, a convention is generally adopted to only count trees with a diameter at breast height that is above some minimal value (e.g. 10 mm or 10 cm, depending on the site). As discussed in Section 7.1, as long as *some* convention is consistently adhered to within any given dataset, the MaxEnt procedure can be applied and tested.

Another aspect of censusing that can lead to trouble involves the temporal dimension. If you keep returning to a plot, day after day, year after year, you will see individuals, and probably species, coming and going. Should METE be construed as a theory about an ecosystem at a slice in time, or should S_0 be the number of species that accumulate over a lengthy time period? If the latter, should the state variable N_0 be the sum of the number of distinct individuals that are ever in the system over the entire time? Similarly would the state variable E_0 be the cumulative metabolic rate over the time period? Ultimately, as with many other questions about application of MaxEnt, including the way we defined the fundamental quantity $R(n, \epsilon)$ and our choice of state variables (why energy, why not water?), this must be answered empirically. While we opt for the slice-in-time interpretation here, future investigation of species–time relationships (Adler et al., 2005; White et al., 2006; Carey et al., 2006, 2007) within a MaxEnt formalism could prove interesting.

Finally, to repeat the point made in Section 7.1, what we call a species does not matter, provided we use consistent criteria to define the units of analysis.

8.3 The species-level spatial abundance distribution

Examining the predicted species-level spatial abundance distribution $\Pi(n|A, n_0, A_0)$, we encounter what is among the most surprising predictions of METE. This concerns the distribution of the individuals within a species across the two halves of a plot, or $A = A_0/2$. As implied by Eq. 7.51, METE predicts that all allocations across the two halves of a plot are equally likely.

Equation 7.51 looks somewhat preposterous as an ecological prediction. Is it really possible that, if there are n_0 individuals in a plot of area A_0, then the $n_0 + 1$ allocations $(n \mid n_0 - n)$ are all equally likely? To test the prediction, we can examine, for each species, the fraction of its individuals that are in the left-hand half of the plot, and we can also examine the fraction of its individuals that are in the upper-half of the plot. If Eq. 7.51 is correct, a plot of all these fractions should reveal that all fractions are equally likely.

A convenient way to portray these fractions is with rank–fraction graphs, which avoid the problems with binning data that we discussed in Section 3.4. If, for all species with $n_0 > 1$ in A_0, the fraction of individuals in the left- or top-half is rank-ordered, with rank 1 corresponding to a fraction equal to 1 and the highest rank corresponding to a fraction equal to 0, then if Eq. 7.51 is correct, a graph of observed fraction versus rank, in which each data point is a species, should be a straight line, declining from 1 at rank 1 to 0 at the highest rank. To see this, recall the discussion in Section 3.4.6; a straight-line rank–fraction graph informs us that the probability distribution of the fractions is a constant, which is the equal allocation prediction.

In contrast, if the distribution of individuals is, for example, randomly chosen from a binomial distribution, then species will be most likely to have roughly equal numbers of individuals in the two halves of the plot (see Section 4.1.1) and most of the data points on a rank–fraction graph will occur with fraction $= 0.5$, giving, not a straight descending line, but rather a nearly horizontal line at fraction $= ½$, over most rank values, with an upturn at low rank and downturn at high rank.

Figure 8.1a–e show such rank–fraction graphs for each of a left–right and top–bottom bisection of A_0 for five datasets with spatially-explicit plant data. The random placement model (Coleman, 1981) prediction (see Section 4.1.1) is also shown on the graphs. The data generally support the equal-allocation prediction of METE, although we note that the 50-ha tropical forest plot at Barro Colorado Island (BCI) deviates the most from theory.

The iterated version of METE, which is equivalent to the HEAP model, predicts this pattern will hold at any spatial scale in the following sense. If a species is found

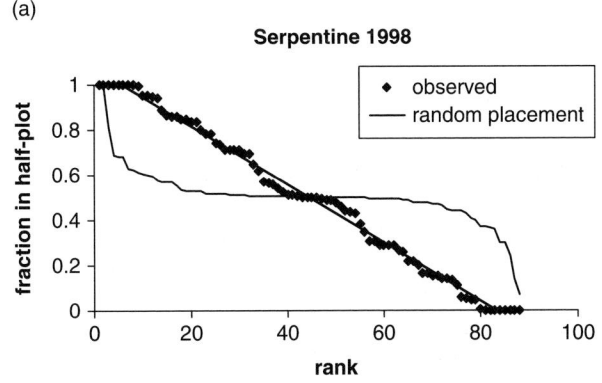

(b)

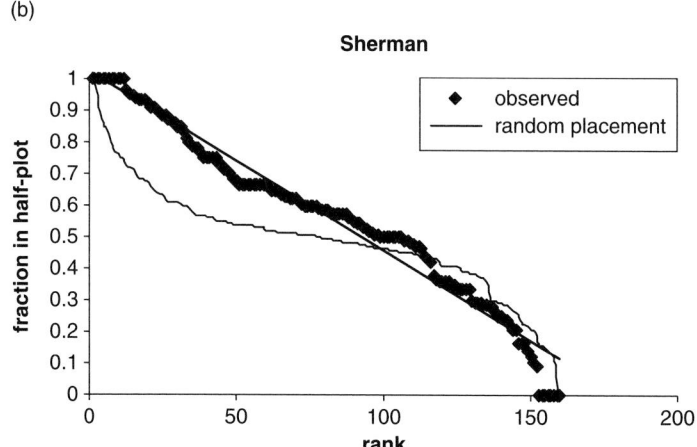

(c)

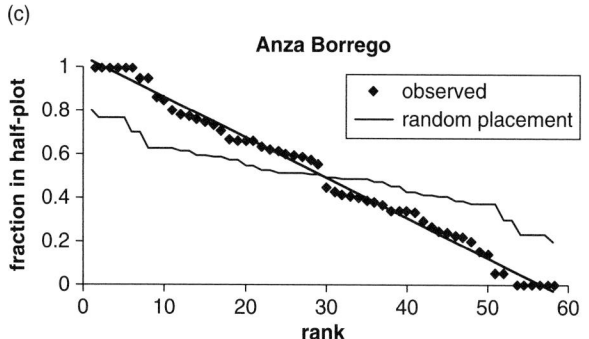

(d)

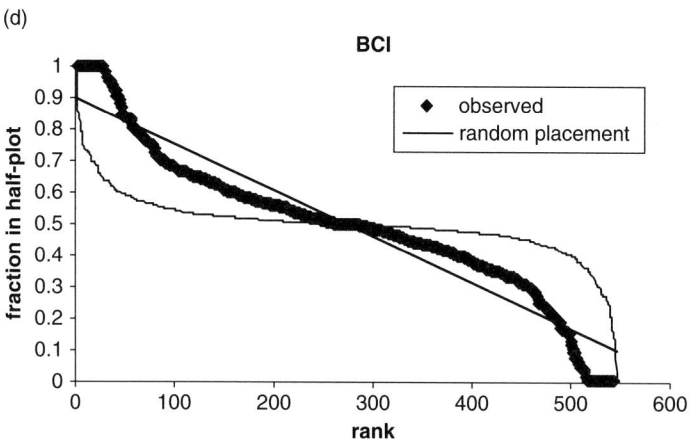

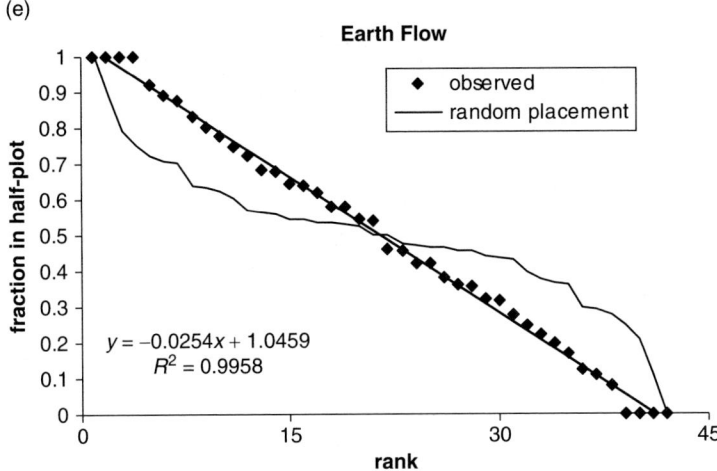

Figure 8.1a–e Comparison of data with the prediction that individuals within a species are uniformly distributed between two halves of a plot (Eq. 7.51) for the five sites labeled in the figures. The METE prediction is a straight line extending from a fraction ≈ 1 at the lowest rank to a fraction ≈ 0 at the highest rank. The random placement model prediction shown in the figure is the binomial distribution, and the straight line is a linear fit to the data.

in, say, the top-half of one of the plots shown in the figure, and has abundance n_1, in that half plot, then all n_1+1 allocations of its individuals between the top-left quadrant and the top-right quadrant should be equally likely. Thus, if the fractions of each of the species in the top-half of A_0 that are in the upper-left quadrant are rank-ordered and plotted again as a rank–fraction graph, the same straight line should result. This result is predicted to hold for bisections of subplots within A_0 down to any scale. Because the actual sizes of locations of the full-scale plots of area A_0 in our test sites are, for the most part, randomly selected out of the landscape, we have no reason to think that there was something privileged about a test carried out at the scale of a bisection of A_0. But we strongly emphasize that finer-scale tests have only been carried out for a few datasets. Conlisk et al. (in preparation) show further tests of the bisection prediction and conclude that at finer scales than examined here, the prediction fares less well in some cases. Clearly, further examination of the conditions under which the prediction fails is needed. Exercise 8.1 asks the reader to take the serpentine dataset in Appendix A and conduct tests of Eq. 7.51, analogous to those shown in Figure 8.1, at finer scales than simply the first bisection.

Consider, next, $\Pi(n|A, n_0, A_0)$ for arbitrary values of A. We are now interested in the predicted distribution of abundances in a cell of area A, if the *only* prior information is that the species has n_0 individuals in A_0. Note that this is less information than we would have if we also knew it had some specified number, n_1, of individuals in the left-half of A_0. Within the framework of MaxEnt, this makes a huge difference; as discussed in Section 4.1.2.2, the HEAP distribution results if we iterate the equal allocation rule at every bisection, whereas if we only know the abundance at scale A_0, then the bounded Negative Binomial Distribution results. The difference between these two models is indicated in Figure 4.1a and b by the lines labeled "neg binom $k = 1$," and "HEAP."

One method for testing the MaxEnt cell-occupancy predictions, $\Pi(n)$, consists of comparing the observed abundances of species at large scale with presence/absence data from small cells. This has practical implications in conservation biology because often only presence–absence data are available and knowledge of species–abundance is sought. Figure 8.2, from Harte et al. (2008), shows that of the 782 plant species, from six of the sites (BCI, Cocoli, Luuillo, San Emilio, Sherman, Serpentine, 1998) in Table 8.1 that were tested, MaxEnt appears to adequately predict abundance from the fraction of occupied, cells.

Another test of the theory consists of separately rank-ordering the cell occupancy values that are predicted and those that are observed, and then plotting the predicted versus observed values. A slope of 1 and $R^2 = 1$ would indicate perfect agreement. Doing this for each of the 14 species with $n_0 > 30$ from the serpentine 1998 census yields a mean slope of 0.87, with a standard deviation of 0.37. The R^2 value averaged over the 14 graphs is 0.88 and the lowest R^2 is 0.77.

Finally, to convey graphically the differences between different model predictions Figures 8.3 and 8.4 compare non-iterated MaxEnt, HEAP, and the Coleman

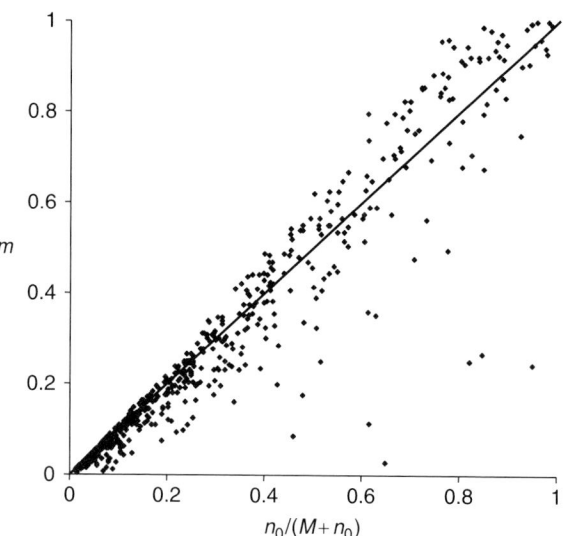

Figure 8.2 Cell occupancy predicts species–abundance. Observed fraction (m) of occupied cells of area A are plotted against MaxEnt predictions ($n_0/(M+ n_0)$), where n_0 is the species–abundance in A_0 and $M = A_0/A$. The graph shows 782 plant species from six sites. The one-to-one line of perfect agreement is also shown; from Harte et al. (2008).

The species-level spatial abundance distribution • 185

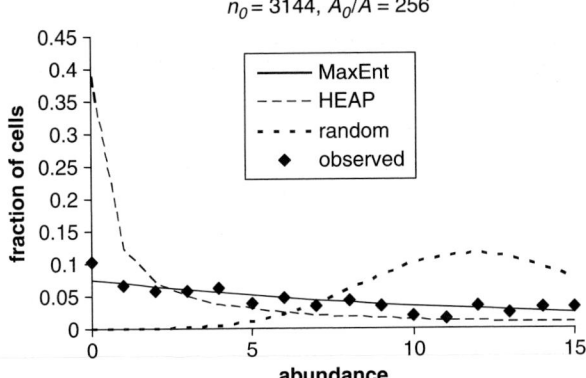

Figure 8.3 Observed species-level spatial abundance distributions and predictions from MaxEnt, HEAP (Harte et al., 2005), and the random placement model (Coleman, 1981). Representative distributions are shown for two species in the BCI plot at a spatial scale $A = A_0/256$, $n_0 = 638$.

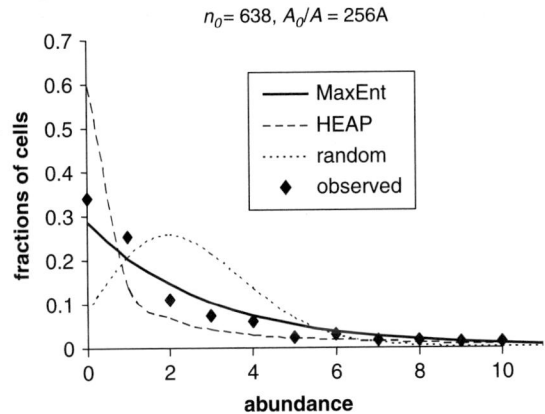

Figure 8.4 Observed species-level spatial abundance distributions and predictions from MaxEnt, HEAP (Harte et al., 2005), and the random placement model (Coleman, 1981). Representative distributions are shown for two species in the BCI plot at a spatial scale $A = A_0/256$, $n_0 = 3144$.

(1981) random placement model with two species from the BCI plot. Note that the data for $n_0 = 638$ tend to fall midway between the HEAP and the MaxEnt prediction, while the more abundant species ($n_0 = 3144$) tends to follow the non-iterated MaxEnt prediction more closely. In Harte et al. (2008) this was shown to be a fairly general result for the plant species tested there.

To summarize, agreement of census data with the METE prediction for the species-level spatial abundance distributions is substantial but not overwhelming. While data from numerous sites sets do appear to obey the equal allocation prediction (Eq. 7.51), some sites, such as BCI, show sizeable deviations from the prediction. The predicted relationship between cell occupancy and species–abundance (Fig. 8.2) exhibits scatter around the line of perfect agreement, but generally good agreement.

8.3.1 A note on use of an alternative entropy measure

METE is based upon the use of the Shannon information entropy measure, both to predict the form of $\Pi(n)$ and $R(n, \epsilon)$. If a different measure, as discussed in Box 5.2, is used, quite different predictions can emerge. For example, numerical evaluation of the MaxEnt solution for $\Pi(n)$, using the Tjallis entropy measure (Eq. 5.23), yields distributions that do not resemble observed spatial abundance distributions unless the parameter, q, that defines the family of Tjallis entropies is taken to be very close to $q = 1$. For $q > 1$, approximate power-law dependence on n is predicted, while in the limit of $q \to 1$, the Tjallis entropy approaches the Shannon entropy. I do not see a compelling reason other than curiosity for further exploration of the Tjallis or other non-Shannon measures of entropy in macroecology.

8.4 The community-level species–abundance distribution

For reasons discussed in Section 3.4, rank–variable graphs often provide a useful way to compare predicted probability distributions with data. Figure 8.5a–h compare the METE prediction of a logseries distribution (Eq. 7.32) with data from representative sites. The predicted rank–abundance distribution, plotted as log(abundance) versus rank, is a straight line for large rank, but with an upturn at low rank. The strength of the predicted upturn varies from barely detectable (e.g. the 1998 serpentine plot) to quite dramatic (Cocoli). The reason the strength of the upturn varies is that the measured value of the ratio, S_0/N_0, varies from site to site. The larger that ratio, the larger is β, the combination of Lagrange multipliers that appears in the exponent of the predicted log series distribution. And the larger β is, the larger is the upturn on a log(abundance) vs.rank graph. Indeed, if $\beta = 0$, a graph of log(abundance) versus rank would be a straight line.

Agreement between the METE prediction and the data shown in the graphs is generally good, although we note here two instances of significant deviation. One is for the BCI data (Figure 8.5b), and the other is for one of the Hawaiian arthropod datasets (site KH in Figure 8.5h). Comparison with Fig. 3.5c shows that the BCI abundance distribution appears to be intermediate between a lognormal and a logseries. The Hawaiian KH abundance distribution does not appear to be well-described by any of the functions shown in Fig.3.5. We will return to this point in

(a)
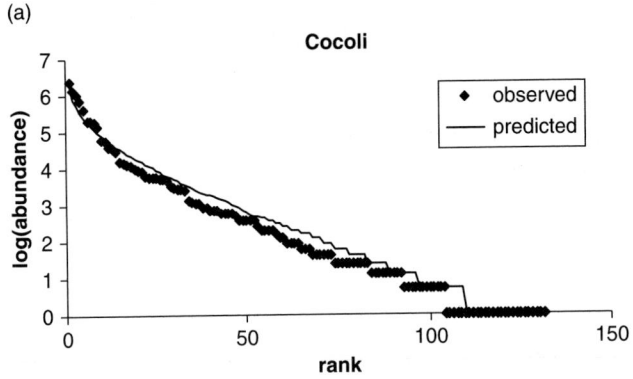

(b)
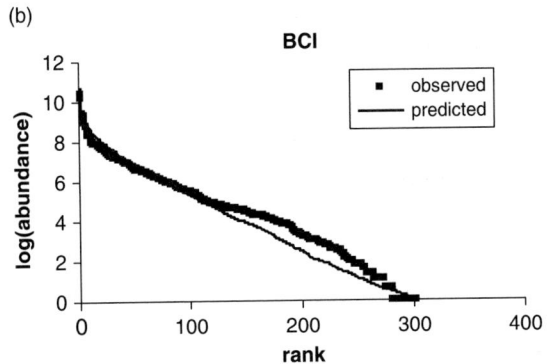

(c)
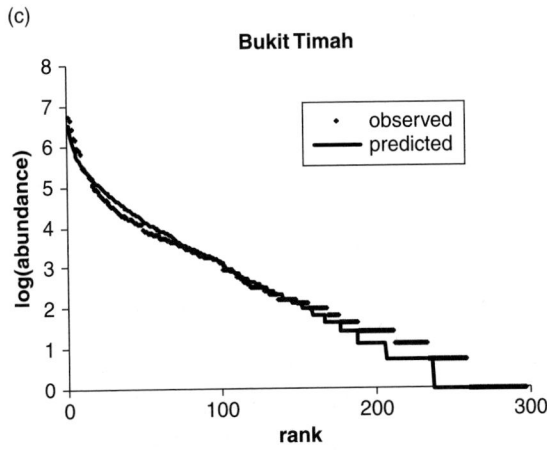

188 • *Testing METE*

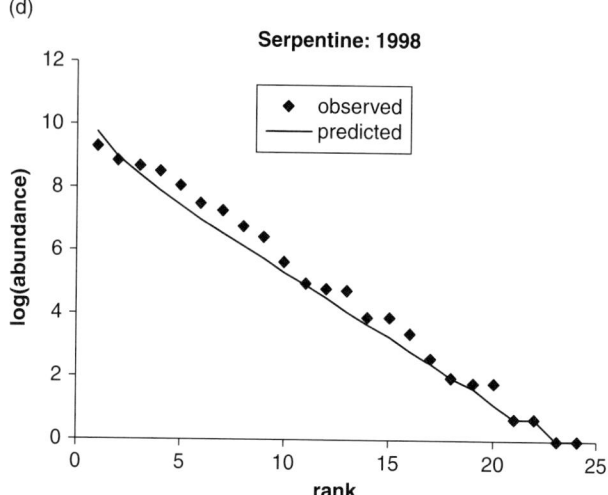

(d)

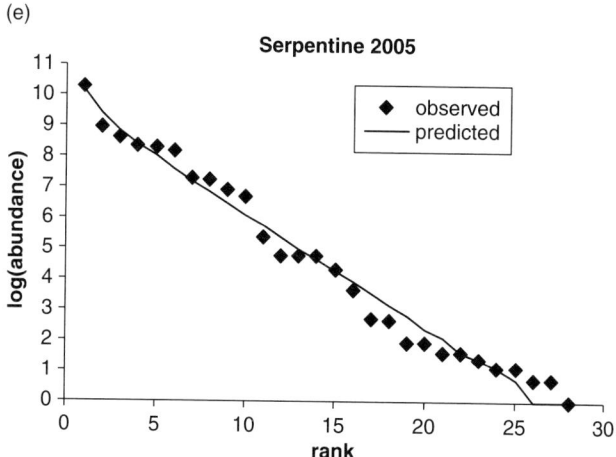

(e)

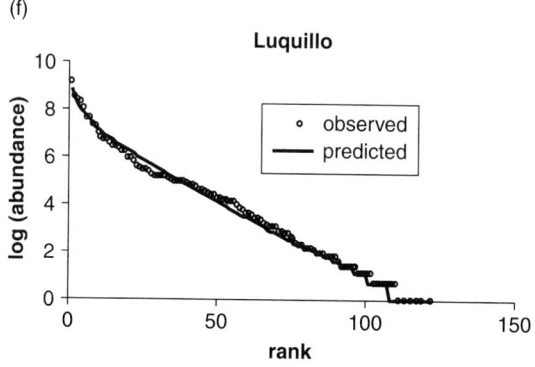

(f)

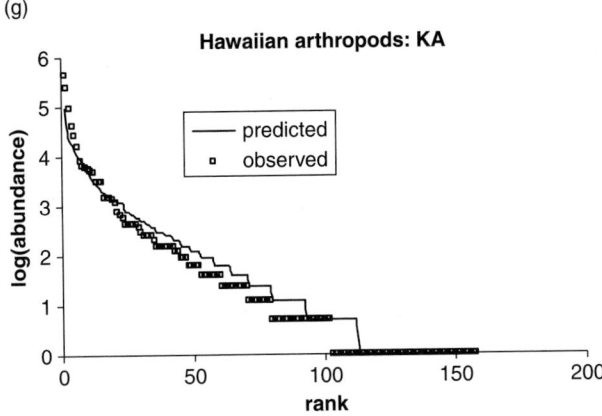

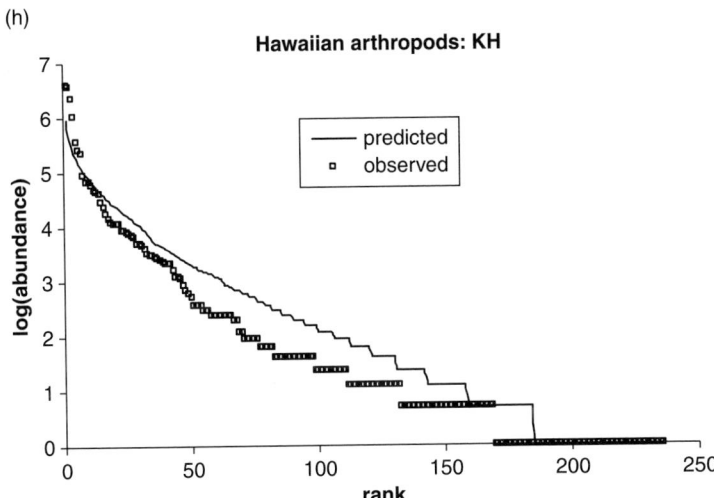

Figure 8.5 Comparison of data with the METE prediction for log(abundance) versus rank for sites labeled in graphs a–h.

Section 8.7. Harte et al. (2008) present additional tests of the METE prediction for the SAD, using data from other tropical datasets (Sherman and San Emilio in Table 8.1) and show that METE explains over 97% of the variance in the abundance distribution for the six sites tested there.

In conservation biology, one of the important things to know about an ecosystem is the number of rare species, for these are the ones likely to be most at risk of extinction. As shown in Eq. 7.40 and the discussion preceding it, the predicted number of species with $n_0 <$ some fixed number m, is $(1+ 1/2 + 1/3 +\ldots+ 1/m)\beta_0 N_0$, provided $\beta_0 m \ll 1$. Figures 8.6 shows a test of the ability of METE to predict the number of rare species, taken here to be 10 or fewer individuals ($m = 10$), for all of our sites in Table 8.1 with abundance data. METE clearly predicts the numbers of rare species accurately.

METE also predicts the abundance of the most abundant species in an ecosystem (Eq. 7.41). Figure 8.7 provides a test for the same sites as in Figure 8.6; again, most of the variance in the data is captured by the prediction.

8.5 The species–area and endemics–area relationships

As discussed following Eq. 7.69, another remarkable prediction of METE is that all species–area curves collapse onto a universal curve if the data are appropriately rescaled. More specifically, METE asserts that at any scale, A, the local slope of the SAR at that scale (given by the tangent to the curve when plotted as $\log(S)$ versus $\log(A)$) is a universal function of the ratio $N(A)/S(A)$. This is tested in Figure 8.8, where data from 41 empirical species–area curves are plotted. Although there is scatter around the predicted line of collapse, the overall agreement with the prediction is good. Further tests of METE prediction for the SAR are provided in Harte et al. (2008).

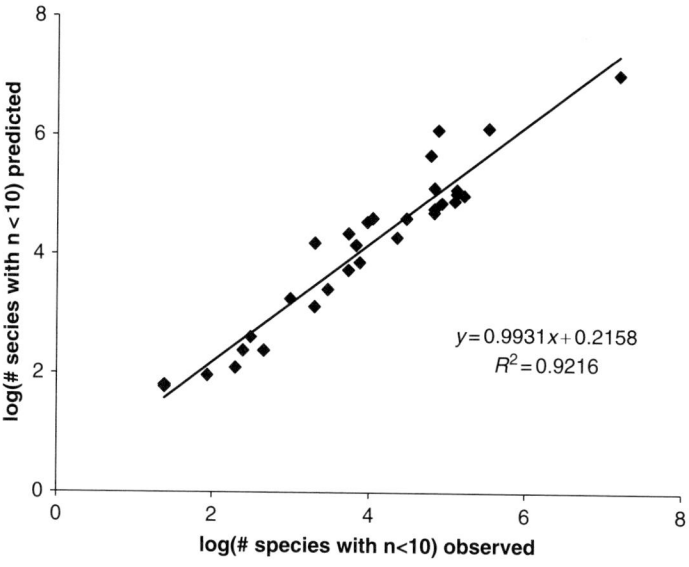

Figure 8.6 Test of the METE prediction (Eq. 7.40) for the number of rare species (here taken to be $n \leq 10$) for all the sites listed in Table 8.1. Only the values of N_0 and S_0 are used to make the predictions.

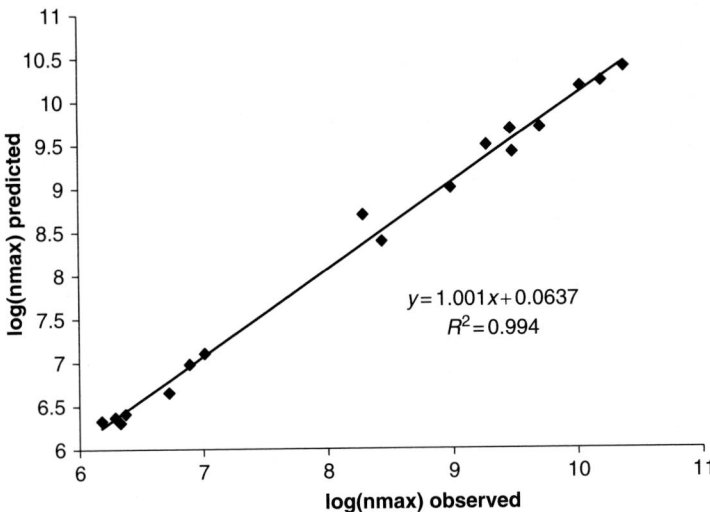

Figure 8.7 Test of the METE prediction (Eq. 7.41) for the abundance of the most abundant species at each of the sites in Table 8.1. Only the values of N_0 and S_0 are used to make the predictions.

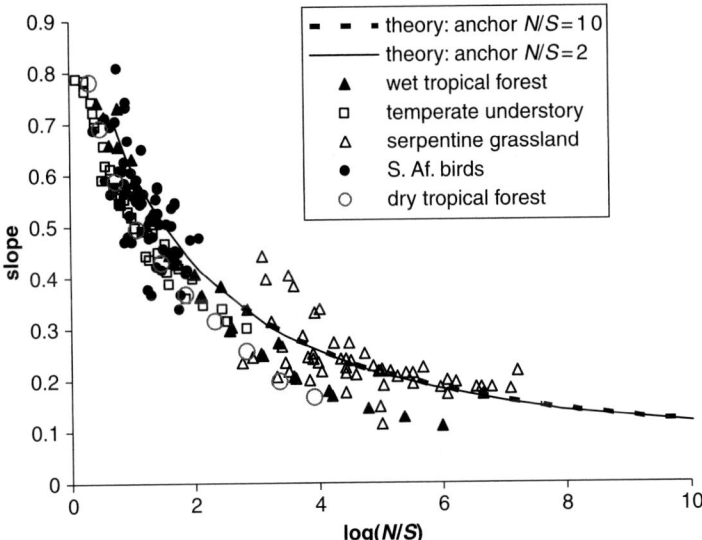

Figure 8.8 Test of scale collapse predicted by METE. Predicted and observed values for the scale-dependent SAR slope parameter, z, plotted as a function of $\log(N/S)$. The dashed and solid lines correspond to the theoretical predictions based on anchor-scale values of $N_0 = 100$, $S_0 = 10$, and $N_0 = 40$, $S_0 = 20$, respectively. Two predictions for the shape of the universal z versus $\log(N/S)$ curve are included simply to show that the curve is independent of the anchor scale values of the state variables. The "wet tropical forest" data are from the BCI, Cocoli, and Sherman in Table 8.1; "serpentine grassland" include both datasets in Table 8.1; the "dry tropical forest" data are from the San Emilio dataset in Table 8.1; and the "temperate understory" data are from the subalpine forest understory flora in Table 8.1.

Two recent empirical papers (Green et al., 2004; Horner-Devine et al., 2004) describing SARs for microorganisms conclude that, when fitted with a power-law function, very low z-values (~ 0.05) result. At least qualitatively, this is consistent with our predicted universal SAR because the N/S values for their datasets are far larger than for typical microorganism census data. In other words, the z-values for these SARs are way out on the far tail of the universal curve where N/S is large and z is small.

The data in Figure 8.8 are all from situations where at some relatively large scale, A_0, the state variables are known, at smaller scales, species' richness is also known, and METE is used to predict species' richness at smaller scales. But of more practical interest is the use of METE to upscale species' richness from small plots, where data are available, to much larger scales. And for this purpose, it is possible that the discrepancies between predicted and observed slopes, z, of the SAR that are evident in Figure 8.8 could result in accumulating errors that give markedly erroneous results for large-scale species' richness.

To test the reliability of upscaling with METE, we estimated tree species' richness, for trees of at least 10 cm dbh, in the entire Western Ghats preserve (Western

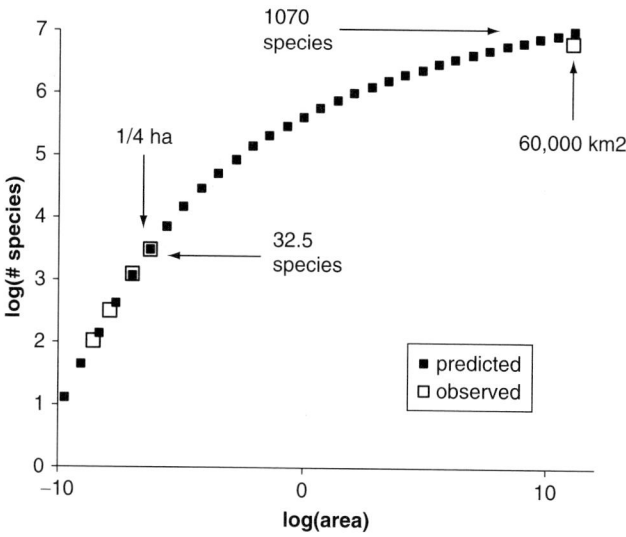

Figure 8.9 Test of upscaling and downscaling predictions for tree species' richness in the Western Ghats (I) dataset. The input information for all the predicted values of species' richness shown in the graph are the averages of total abundance and species' richness in 48 ¼-ha plots scattered throughout the 60,000 km² preserve (Krishnamani et al. 2004). In those small plots, abundance and spatial distribution data for every tree species with ≥ 30 mm dbh were obtained and used to construct 48 species–area relationships spanning nested areas within each plot of 0.25, 0.125, 0.05, and 0.025 ha. Area is in units of square kilometre.

Ghats (I) in Table 8.1). The anchor-scale census data are from 48 ¼-ha plots scattered throughout the 60,000 km² preserve. Figure 8.9 shows the results of the upscaling test carried out by repeated interation of Eqs 7.70 and 7.71, from the knowledge of the state variables at the anchor scale of ¼ ha. The theory predicts species' richness all the way up to the entire 60,000 km² of the preserve. The figure also shows that the downscaling SAR prediction of METE also works well at that site; theory predicts the species' richness at scales < ¼ ha from the state variables at ¼ ha scale.

Another, independent, dataset from the Western Ghats contains tree census information from 96 1-ha plots scattered throughout 20,000 km² comprising roughly the middle-third of the preserve in the north–south direction (Western Ghats (II) in Table 8.1). Our upscaling procedure (Eqs 7.70 and 7.71) predicts a total of 544 tree species; the current estimated value for that portion of the Preserve is 580 (D. Vanugopal, pers. comm.).

The agreement between the predicted endemics–area relationship (Section 7.7) and observation is also remarkably good. Two examples are shown in Figure 8.10a and b.

8.6 The distribution of metabolic rates

METE predicts the distribution of metabolic rates across all individuals, $\Psi(\epsilon)$, the distribution of metabolic rates across all the individuals within a species of known abundance, $\Theta(\epsilon|n)$, and the distribution across species of the average metabolic rate

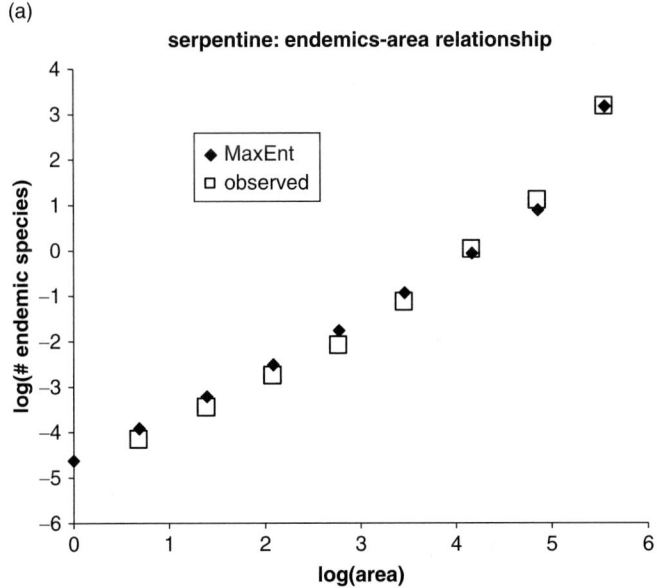

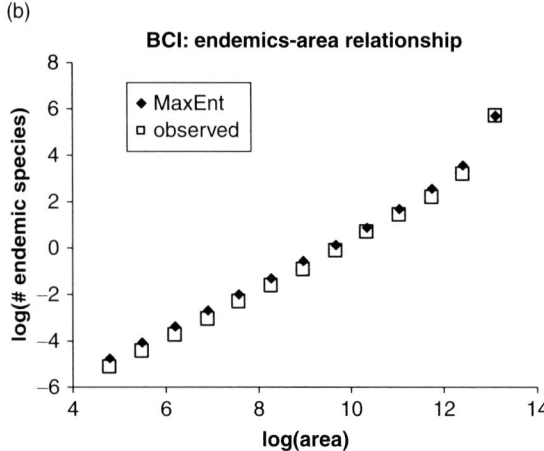

Figure 8.10 Comparisons of the predicted endemics-area relationship (Eq. 7.74) with data from: (a) the serpentine grassland 1998 census; (b) the BCI 50-ha plot.

of the individuals within species, $v(\epsilon)$. Testing these predictions is not straightforward, however, because metabolic rates of each individual, and thus evaluation of the total metabolic rate, E_0, are rarely measured directly in the kinds of censuses that also result in values of our other state variables, S_0 and N_0. To test the predictions of metabolic rate distributions, we have to invoke one or both of two assumptions: a metabolic scaling rule and an allometry.

If we assume the validity of the metabolic scaling rule, $\epsilon \sim m^{3/4}$, then we can estimate metabolic rates from body mass measurements, which are easier to obtain than are measurements of metabolic rates. But there is evidence that the ¾ power scaling rule is not strictly valid across the whole range of body mass variation in nature. For trees, especially, it is not clear what the correct quantity is to use for individual mass. Moreover, in many cases, particularly with plants, direct measurements of body mass are not available. Often what is measured for trees is basal area or diameter at breast height from which body mass can be estimated if the plant allometry is understood. But such allometries are controversial (Muller Landau et al., 2006). Simple rules can be invoked, such as that an individual plant's metabolic rate is proportional to its total leaf area, but the body of empirical work on plant allometry reveals significant exceptions.

Given these data constraints, we focus here on just a few tests of METE's metabolic rate predictions. Figure 8.11a and b show fairly good agreement between the METE prediction and available data for two datasets. On the other hand, Figure 8.11c shows that the Hawaiian site, KH in Table 8.1, for which the abundance prediction differed from observation, also shows poorer agreement with the METE prediction for the metabolic rate distribution. We discuss the implications of this and other discrepancies in Section 8.7.

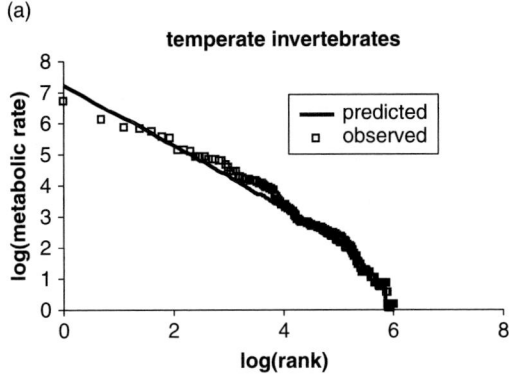

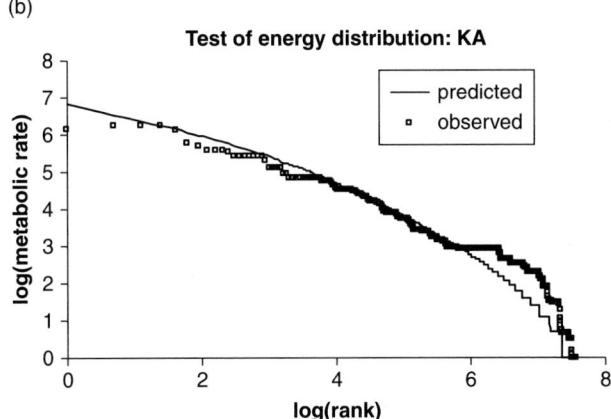

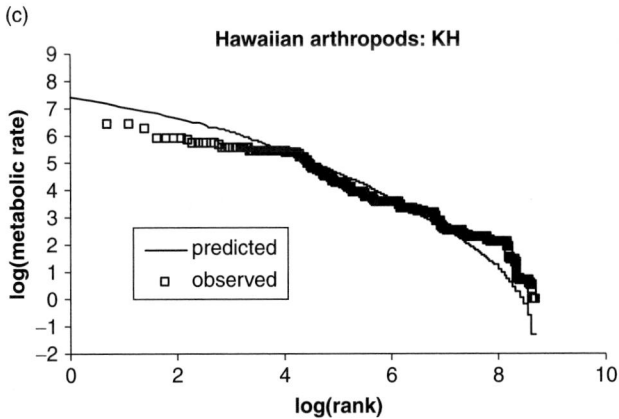

Figure 8.11 Comparisons of the predicted distribution of metabolic rates across all individuals in the community, $\Psi(\epsilon)$, with data from: (a) a pit-trap census of temperate invertebrates; (b) Hawaiian arthropods (KA in Table 8.1); (c) Hawaiian arthropods (KH in Table 8.1). In all three cases, a metabolic scaling relation $\epsilon \sim m^{3/4}$ is assumed.

8.7 Patterns in the failures of METE

Two types of situations are plausibly going to result in failures of METE predictions. First, systems with state variables that are relatively rapidly changing in time might not be well-described by a theory based upon a static conception of state variables. Hence, we might look for failures in systems undergoing rapid diversification, degradation, or succession in the aftermath of disturbance.

Second, systems with very heterogeneous habitat characteristics over large spatial scales are not likely to satisfy the METE species' richness scale-up prediction. In other words, attempting to scale up species' richness using data from small plots, each within a single habitat type, should underestimate species' richness at large scales, where different species lists from the differing habitats all add up to the large-scale species list.

There is evidence from empirical tests of METE that the above generalizations are both in fact valid.

In Section 8.4, the failure of METE to accurately predict the species–abundance distribution at one of the Hawaiian arthropod census sites (KH) was noted, while in Section 8.6, its failure to predict the metabolic rate distribution at that same Hawaiian site was noted. This is particularly interesting because at the other Hawaiian site (KA) plotted, and at three more sites not plotted, METE better captures the central tendencies in both the abundance and metabolic rate distributions. Is this a coincidence or is it indicative of something more interesting—a systematic pattern of failure that may provide a hint as to how to improve the theory? Figure 8.12 may provide a hint of the answer to that.

The site that rather dramatically fails to obey the METE predictions, KH, is the site of intermediate age. Perhaps that site is one at which species' diversification is

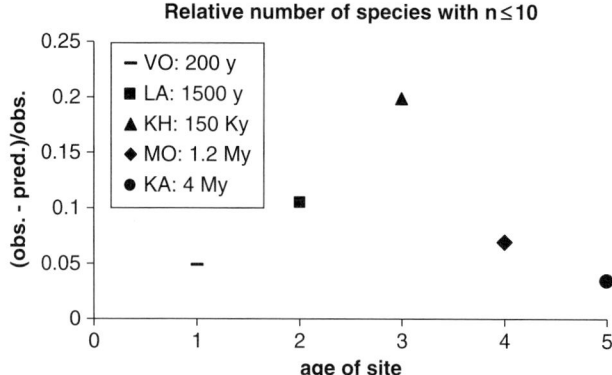

Figure 8.12 Relationship between age of site and discrepancy between predicted and observed number of rare species at the five arthropod census sites in Hawaii listed in Table 8.1.

occurring most rapidly, as might be the case if species' richness follows a logistic-type curve starting with the emergence of a colonizable initially barren site?

Going beyond such speculation, however, there is evidence from censuses of moths at a variety of plots at Rothhamstead (Kempton and Taylor, 1974) that the logseries species–abundance distribution generally better describes the data except for plots that are known to be undergoing rapid change in the aftermath of disturbance. A study by E. Newman now underway at a site in Colorado that is recovering from a large erosion event nearly 100 years, should provide further insight into the pattern of deviation of the macroecological metrics from the METE predictions under conditions of relatively rapid ecological change.

A failure of METE has also has been demonstrated when it is used to upscale species' richness in heterogeneous regions. Kunin et al. (in preparation) show that METE hugely underestimates total plant species' richness in the entirety of the UK, if the average S and N from numerous 0.02-ha plots are used as input data. Thus the second of our two suggested scenarios for the failure of METE is supported. This is hardly suprising, for the plots span a wide variety of very distinct habitats, with nearly non-overlapping species lists.

The first step that is needed to extend the validity of METE's predictions for abundance and metabolic rate distributions to systems in which change in state variables is relatively rapid, is to develop an understanding of the dynamics of the state variables. An approach to doing this is described in Section 11.3. The first step needed to extend the validity of the METE upscaling procedure to environmentally heterogeneous regions is to learn how to incorporate data on spatial correlations into the theory; this would allow use of species-level commonality predictions to improve the upscaling strategy described in Section 7.5. An approach to doing that is described in Section 11.1.

8.8 Exercises

*Exercise 8.1

Using the 1998 serpentine date in Appendix A, test Eq. 7.51 at finer scales than the first bisection.

*Exercise 8.2

Calculate the METE prediction for the collector's curves for the BCI and the 1998 serpentine plot data and compare your result with the graphs in Figure 3.9. To do the calculation you will need to use the results in Section 7.7 and the data from Table 8.1.

Part V

A wider perspective

9

Applications to conservation

Efforts to estimate the biological consequences of human activities often can be broken down in to a set of narrower, or at least more specific, research questions that conservation biologists grapple with every day. They include the following:

- *How can we estimate species' diversity at large scales from small-scale census data?*
- *How can we infer abundance from sparse presence/absence data?*
- *How can we estimate the number of species that will be lost under habitat loss?*
- *How can we best determine the most likely associations of habitat characteristics with species presence?*

Here I briefly summarize ways in which METE can contribute to answering these questions.

9.1 Scaling up species' richness

The first item on this list has long pre-occupied ecologists. Are there 2 million or 50 million species on the planet? How many species of beetles inhabit the Amazon? (May, 1990). Knowledge of the shape of the SAR from spatial scales ranging from that of small plots to entire biomes would allow estimation of biome-scale species' richness from small plot data; and METE provides exactly that. The theory predicts that the slope z at any scale, A, within a biome is a universal, decreasing function of the ratio $N(A)/S(A)$, where $N(A)$ is the total number of individuals all the species at scale A, and $S(A)$ is the number of species at that scale. While more testing is needed, particularly with animal data, the theoretical prediction is in good agreement with observations for a wide range of habitats and spatial scales ranging from plots of order several square meters to more than 60,000 km^2 (Figures 8.8 and 8.9).

There are many situations where this approach will clearly fail. As discussed in Section 8.7, the method cannot possibly predict from small plot data the species' richness in regions containing very diverse habitats. To extend the theory to heterogeneous biomes, it will be necessary to incorporate information about species' turnover as a function of distance between small plots. An approach to doing that with MaxEnt is discussed in Chapter 11.

The current theory, however, in conjunction with tree-canopy fumigation data on arboreal arthropod diversity, could be used to upscale arthropod species' richness from the scale of individual tree canopys to that of a larger region of relatively homogeneous habitat.

9.2 Inferring abundance from presence–absence data

Inferring abundance from presence–absence data, or from the results of other incomplete census designs, is vitally important in conservation biology for reasons described in Section 3.5.3. For that reason, several authors have developed approaches based on various models of spatial structure (He and Gaston, 2000; Conlisk et al., 2007a). The essential idea behind any approach to this task is the following. For any assumed species-level spatial abundance distribution, $\Pi(n|A, n_0, A_0)$, that depends uniquely on n_0, a known value of Π at any value of n determines n_0. In particular, with presence–absence data, a measured estimate for the fraction of cells of area A in which the species is absent, $\Pi(0|A, n_0, A_0)$, suffices to determine n_0. A test of the METE prediction of n_0 values from plant-occurrence data was shown in Figure 8.2. Further testing, particularly with animal data, is clearly needed. Moreover, it would be useful to work out how the accuracy of the prediction is improved as either more plots or larger plots (of area A) are censused for presence–absence, and similarly how accuracy improves if the actual abundance distribution versus just presence–absence in the small plots is measured. And just as we saw was the case with upscaling species' richness, incorporating knowledge of the spatial correlation structure of populations would likely improve the accuracy of abundance estimations based on presence data.

9.3 Estimating extinction under habitat loss

In a landmark paper, Thomas et al. (2004) introduced a method for estimating the number of species likely to become extinct under habitat loss or degradation from climate change or land-use practices. They posited that each species obeys a scaling formula for the probability of its persistence when the area of that species' suitable habitat shrinks. The formula they suggested reads:

$$P_{\text{after}} = P_{\text{before}}(A_{\text{after}}/A_{\text{before}})^z, \qquad (9.1)$$

where P_{after} and P_{before} are the probability of survival in the reduced habitat of area A_{after} and the probability of survival in the original habitat of area A_{before}. They assert that z, the likely range for the constant z, is 0.25 to 0.35, and take 0.15 as a "conservative" case, in the sense that it leads to a lower extinction probability. Equation 9.1 was motivated, loosely, by the notion that the SAR is a power-law relationship. Species-level scaling rules, however, have been predicted to vary across species and across spatial scale, and a firm theoretical basis for Eq. 9.1 is lacking.

METE provides a means of accomplishing the same goal as Eq. 9.1, but for the result to be relevant to the concerns of conservation biologists, an explicit assumption has to be made that relates change in population size to change in extinction risk. Assume, for example, that the extinction probability for a species is a function of the ratio $r = n_{after}/n_{before}$ (e.g. extinction occurs if, and only if, r is less than a critical ratio r_c). Here n_{after} and n_{before} are population sizes after and before the loss of habitat. From the predicted form of $\Pi(n|A, n_0, A_0)$, METE then predicts an extinction probability that is a function of the ratio of A_{after} to A_{before}, as well as of n_{before}. The effective z-value for each species is now a known function of the initial abundance n_{before} of the species in an initial area A_{before}, and of the ratio A_{after}/A_{before}; it is not a constant. In particular, for n_{before} and $A_{before}/A_{after} \gg 1$, the probability that a species retains at least a critical fraction r_c of its initial abundance is given by (Kitzes and Harte, in preparation):

$$P\left(\frac{n_{after}}{n_{before}} > r_c\right) = \frac{[n_{before}\kappa/(1+n_{before}\kappa)]^{r_c n_{before}} - [n_{before}\kappa/(1+n_{before}\kappa)]^{n_{before}}}{(1+n_{before}\kappa)\log(1+1/(n_{before}\kappa))}, \quad (9.2)$$

where $\kappa \equiv A_{after}/A_{before}$. Derivation of Eq. 9.2 is left as an Exercise (9.1) for the reader. An analogous expression can be derived for any extinction criterion. For example, if extinction is assumed to occur if $n_{after} - n_{before}$ exceeds a critical value, r_c, then an expression for $P(n_{before}-n_{after}>r_c)$ follows again from the predicted form of $\Pi(n|A, n_0, A_0)$ (Exercise 9.2).

9.4 Inferring associations between habitat characteristics and species occurrence

Suppose we want to fill in incomplete census data for a landscape on which we have some knowledge of an environmental variable that we believe influences the likelihood of local occurrence of a species. We start with spatially explicit knowledge of the value of the environmental variable over the landscape, and, in addition, we have census data over some portion of the landscape. How can we use MaxEnt to improve our knowledge? A considerable literature exists on this topic (see, for example, Phillips et al., 2004, 2006; Elith et al. 2006; Phillips and Dudik, 2008; Elith and Leathwick, 2009) and the reader is encouraged to peruse those works. The explanation provided below is really intended for pedagogic purposes, providing the reader with a simple explanation of the fundamentals behind the strategy of using MaxEnt to infer habitat–occurrence associations.

Consider a landscape with 10 cells and a single species of interest. For each cell, a single environmental parameter, T, perhaps average annual temperature, is measured. We assume, for the sake of simplicity, that it can take on integer values from 1 to 6. In addition, on 6 of the cells, censusing has been carried out and has provided both presence and absence data for the species; that is, on each of those cells it is known whether the species is present or absent. The number of censused cells need not have any relationship to the number of possible T-values. The problem is to determine the probability of presence on the un-censused cells, as a function of

Table 9.1 Made-up values for ten spatial cells of an environmental parameter, T, and for the status of a species occurrence.

Environmental variable, T	Species' status
1	0
2	1
3	—
6	1
4	—
5	0
3	—
5	0
2	—
1	1

1 = present, 0 = absent, — = no data.

the value of the environmental variable for the cell. The available data are shown in Table 9.1. Under species' status, 1 means present, 0 means absent, and – means the cell is not censused.

The first step is to use MaxEnt to derive the probability distribution $P(T|1)$. The notation is the same as used throughout the book for conditional probabilities: P is the probability of finding the environmental variable T in a cell, given that the species is found in the cell. So our study set is the set of three cells in which the species' status is 1. The values of T are all in the range from 1 to 6, so one way we could proceed is to treat this problem like a six-sided die problem. The observed mean value of T is $(2+6+1)/3 = 3$. Following the procedure described in Chapter 6, and calling λ the Lagrange multiplier for this problem, we obtain the following equation for $x = e^{-\lambda}$:

$$x + 2x^2 + 3x^3 + 4x^4 + 5x^5 + 6x^6 = 3 \cdot (x + x^2 + x^3 + x^4 + x^5 + x^6). \quad (9.3)$$

This can be solved numerically giving $x = 0.8398\ldots$ and Z, the partition function, $= 3.4033\ldots$.

The inferred probability distribution is then:

$$P(T|1) = \frac{(0.8398)^T}{3.4033}. \quad (9.4)$$

The next step is to convert this into estimates for the quantities we really want to know: the probabilities that the species is present in the cells that were not censused. For each of those cells, we know the value of T, and so our next step is to derive $P(1|T)$. Here, as above, the value of 1 to the left of the conditionality symbol refers to presence of the species. Consider the quantity $P(A,B)$, which is the probability that two events A and B, occur. We can write this as $P(A, B) = P(A|B)P(B)$, and we can also write it as $P(B|A)P(A)$. Equating these two ways of writing the same thing, we get:

$$P(A|B)P(B) = P(B|A)P(A). \tag{9.5}$$

Rearranging this equation, we get Bayes law in the form:

$$P(A|B) = P(B|A)P(A)/P(B). \tag{9.6}$$

Now let A be the event of presence and B be the event that a value of T is obtained for the environmental variable:

$$P(1|T) = P(T|1)P(1)/P(T). \tag{9.7}$$

To proceed, we can substitute Eq. 9.4 for the first term on the right-hand side of Eq. 9.7. Moreover, $P(1)$ can be estimated from the data in Table 9.1: of the six censused cells, three contained the species. So, based just on available data in the table, our best inference of $P(1)$, the probability of a presence in the absence of any environmental information, is $P(1) = 0.5$. Finally, we could either assume all values of T between 1 and 6 are equally likely, so that $P(T) = 1/6$, or we could take measured values of T and derive a more accurate estimate. Doing the former (but see Exercise 9.3), and putting this together, we arrive at:

$$P(1|T) = \frac{(0.8398)^T}{3.4033} \cdot 0.5 \frac{1}{\frac{1}{6}} = 0.881 \cdot (0.8398)^T, \tag{9.8}$$

and for each T-value it yields our inferred probability of presence.

Filling in the table, for rows 3 and 7, with $T = 3$, we have a probability of presence of 0.52. For row 5, with $T = 4$, we have a probability of presence of 0.44, and for row 9, with $T = 2$, we have a probability of presence of 0.62. If the $P(T)$ estimates are taken from the empirical values in the table, then, for example, $P(3) = 1/5$, and the entry in row 3 would be 0.43, not 0.52.

Aside from allowing the filling in of missing census data at some moment in time, this procedure can also be used to infer climate envelopes that can be used to estimate likely species' ranges under future climate change. For example, suppose we know where, within some landscape, a species is found and not found, and we also know the value of some climate parameter that we think influences the presence of the species across that landscape. We will further assume that species have unlimited ability to migrate or disperse to new locations with suitable climate. We wish to know where the species will be found when the climate parameter changes in the future. Now the missing census data concern the future, not portions of the present, landscape. If the climate conditions in the future can be estimated, then the above procedure allows us to estimate the probability of occurrence in the future at sites that may either overlap or be distinct from the current range of the species.

The numerical example above illustrates just a simple case of what can become a much more complicated situation. For example, we might have not just one, but rather a set of environmentalvariables for each cell. And we might have only presence data, not presence–absence data. Moreover, the approach above made no reference to the explicit spatial structure of the six sites for which presence–absence data were obtained. Hence no information about spatial correlations was used. If spatially explicit data are available, the correlations could be used to provide more information about the likelihood of an association between environmental conditions and species' occurrence probability. We leave it to the reader to consider all these complications and think about how MaxEnt might apply.

9.5 Exercises

Exercise 9.1

Derive Eq. 9.2 from the METE prediction (Section 7.4) for $\Pi(n|A, n_0, A_0)$.

Exercise 9.2

Derive an expression analogous to Eq. 9.2 for $P(n_{before} - n_{after} > r_c)$.

Exercise 9.3

Derive the values of $P(1|T)$ if $P(T)$ is inferred directly from the frequencies of T-values in Table 9.1.

*Exercise 9.4

A frequently cited calculation concludes that if 25% of the area of, say, the Amazon or the Western Ghats is lost to habitat destruction, such as deforestation, then 7% of the species in the original area will disappear from that area. The calculation assumes a power-law SAR with a slope of $z = 0.25$. The general calculation, for arbitrary slope and arbitrary fraction of lost area, works as follows. Let A_0 be the original area and A be the remaining area. Then $S(A)/S(A_0) = (A/A_0)^z$. For the particular case, we get $S(A)/S(A_0) = (0.75)^{0.25} = 0.93$. Hence 7% of the species are lost. Suppose this calculation is repeated but using the predicted SAR from METE. Using Eq. 7.69 for the non-constant value of the slope, z, estimate for the Western Ghats the fraction of species that will be lost if 25% of the area of the Preserve is destroyed. At the scale of 60,000 km^2 you can assume the state variables are $S_0 = 1070$ (predicted by METE; see Figure 8.9) and, because the average number of individuals on the ¼-ha plots is 109, $N_0 = (109)(60.000 \text{ km}^2)/(¼ \text{ ha}) = (109)(6 \times 10^{10} \text{ m}^2)/(2500 \text{ m}^2) = 2.6 \times 10^9$ individuals.

Exercise 9.5

Discuss reasons why the calculation you did in Exercise 9.4 could greatly overestimate and why it could greatly underestimate the actual fraction of species lost as a result of deforestation.

10

Connections to other theories

METE is situated within a constellation of other theories that also attempt to explain a range of patterns in macroecology. Some of these theories are also based on the maximum entropy principle; others are not. Here I sketch some similarities, differences, and synergies between METE and a selection of these other theories.

10.1 METE and the Hubbell neutral theory

Of the various theories of macroecology that I have discussed in this book, arguably the one that is closest in spirit to METE is the Hubbell Neutral Theory of Ecology (NTE; Hubbell, 2001). Here are some points of overlap and difference.

METE is a neutral theory in the sense that, by specifying only state variables at the community level at the outset, no differences are assumed either between individuals within species or between species. Differences between individuals and species do arise in METE because the abundance and metabolic rate distributions are not flat, but these differences emerge from the theory; they are not assumed at the outset. Similarly, in the Hubbell NTE, differences between species arise from stochastic demographics; no a priori differences are assumed.

At a more technical level, additional points of similarity exist. The Hubbell NTE predicts a logseries abundance distribution for the metacommunity; the fractional difference between the intrinsic birth and death rate parameters in the NTE is identical to the parameter β that characterizes the logseries distribution resulting from METE. Related to this, it can readily be shown that Fisher's α (Fisher et al., 1943), a parameter that plays a central role in the NTE, is equal to the product βN_0 in METE, where β is the sum of the two Lagrange multipliers, as desribed in Section 7.2. Another obvious point of similarity is that neither theory assumes a role for either inter-specific interactions or for interactions of organisms and the physical environment. In that sense, both are null theories, and mismatches between data and theory can provide insight into what interactions are actually influential.

The theories differ in important ways, as well. In the Hubbell theory, spatial structure emerges only with the introduction of dispersal kernels for the species (Volkov et al., 2003) and predictions for inter- and intra-specific distributions of metabolic rates or body sizes also require additional assumptions, while in METE they emerge naturally. On the other hand, the Hubbell NTE elegantly relates the speciation rate to the form of the species–abundance distribution; METE in its present form does not (but see Section 11.3).

Why do theories that make such manifestly incorrect assumptions (neutrality, no interactions), as do METE and the Hubbell NTE, work to the extent they do? One rationalization, paradoxically involves the huge number of ways in which species and individuals do differ from one another and do interact with one another and with their environment. Consider the numerous adaptive traits that organisms possess. Adaptations to the myriad combinations of topographic, climatic, edaphic, and chemical conditions that characterize abiotic environments allow species to find many ways to achieve a sufficient level of fitness so that they can persist long enough for us to have censused them. By this argument, trait differences do not influence the shapes of macroecological patterns in nature because nearly all species achieve approximately comparable levels of fitness. This does not mean that systems are in equilibrium, but rather that deviations from equilibrium roughly average out. For further discussion of this see Hubbell (2006).

In this argument, if there were only a small number of significant violations of the neutrality assumptions, then the theories would work worse than if there are many such violations. Because there are so many ways in which species differ, the successes of neutral theories arise for reasons analogous to why statistical mechanics (which despite the name is mechanism-less) work: real molecules are not points objects that collide perfectly elastically, but the gigantic number of ways and times in which violations of those assumptions influence molecular movements average out somehow. For a recent argument that runs counter to this justification of neutral theories, however, see Purves and Turnbull (2010).

It is also possible that the assertion that neutral theories are traitless needs to be revisited. No traits are assumed by METE at the outset of calculations that predict the macroecological metrics. But the theory predicts a distribution of metabolic rates across individuals within and across species, and thus trait differences, which can influence fitness, at the individual and species levels. Moreover, both METE and NTE predict a range of abundances, from very rare to abundant. Abundance, itself, can be considered to be a species trait, although it is a measure of the fitness of the species, rather than of any particular individual, because rare species are at greater risk of extinction than are common ones. So while the theories assume no trait differences, they actually generate them.

10.2 METE and metabolic scaling theories

To a considerable extent, metabolic scaling theories (MSTs) stand independent of METE. MSTs relate the metabolic rates of organisms to their body mass, with the West, Brown, and Enquist (1997) ¾-power scaling theory being the most prominent among these theories. METE, as developed to date, does not predict any particular scaling relationship between metabolic rate and mass. Likewise, MSTs do not predict the various metrics that METE predicts.

The theories are, however, complementary in one sense: METE predicts the conditions under which the various forms (see Table 3.3) of the energy-equivalence, and the related mass-abundance, rule should hold or not hold within ecological communities (see Section 7.9). Recall the motivation behind energy-equivalence: in typical ecosystems we generally see many small organisms and a smaller number of larger organisms. In their various forms, the different energy-equivalence principles all assert that metabolic activity takes place uniformly across the size spectrum. While none of the forms of the energy-equivalence principle, or its close relative the Damuth rule, are strictly derivable from MST, analyses of census data have provided at least limited support in at least some cases (White et al., 2007). To the extent that the energy-equivalence principle is valid, it gives MSTs leverage to better understand how resources are partitioned and ecological communities are structured. For that reason, metabolic scaling theorists have spent considerable effort trying to determine the validity and limitations of the Damuth rule and the energy-equivalence principle. The usefulness of METE's predictions regarding when energy-equivalence *should* hold remains to be tested, but to the extent these predictions are useful, MST will be advanced.

10.3 METE and food web theory

The niche model of food web structure advanced by Williams and Martinez (2000) was briefly mentioned in Section 4.6. It predicts relatively successfully a number of characteristics of food webs, such as the ratio of species (nodes) to trophic linkages, the relative numbers of species found at different trophic levels, and the average number of degrees of separation between pairs of nodes. The model assumes a particular functional form for the distribution of linkages across nodes, the metric we denoted by $\Lambda(l)$ in Table 3.1. Williams (2010) showed that MaxEnt in fact predicts an exponential function for this distribution, as discussed in Section 6.3.4; the beta distribution assumed in the niche model is nearly indistinguishable from the exponential distribution.

Drawing further connections between MaxEnt and network structure may be possible. Section 11.2 suggests one possible future effort in this direction: the derivation of the number of linkages in a web from the METE state variables.

10.4 Other applications of MaxEnt in macroecology

METE is just one of many conceivable applications of the MaxEnt principle to ecology. The underlying logic of MaxEnt is that it is an inference procedure for applying prior knowledge as a constraint on incompletely specified probability distributions. Different theoretical frameworks based on application of the procedure can lead to different outcomes depending on four inter-related choices:

1. The choice of the units of analysis.
2. The choice of the state variables and the independent variables.

3. The choice of the definitions of the probability distributions.
4. The choice of the constraints and the related choice of relative prior distributions.

There is no formal method, no logical inference procedure, for making choices in each of the above categories. Those choices are the art of science, and different theories will differ in some or all of those choices.

For example, we might have used body sizes rather than metabolic rates of individual organisms as an independent variable upon which the fundamental distribution $R(n, \epsilon)$ depended. We then would have used the total mass, M_0, rather than total metabolic rate, E_0, as a state variable. We also might have defined R differently than we did (see definition in Section 7.2.1); for example, we could have defined the fundamental probability distribution as: $R \cdot d\epsilon$ is the probability that, if an individual (rather than a species) is picked from the individuals' (rather than the species') pool, then it belongs to a species with abundance n, and its metabolic energy requirement is in the interval $(\epsilon, \epsilon+d\epsilon)$.

This would lead to a completely different set of predictions! In fact, in developing METE, we explored that option and found it led to unacceptable predictions.

The categorization of individuals within species plays a central role in METE, but there is no reason not to extend the theory to species within genera, genera within families, or even individuals within families. Those extensions remain to be tested; but METE does "stick its neck out" on this point; it should apply at any such level of definition of units of analysis.

At a fundamental level is the choice as to the underlying statistics of the units of analysis. Referring back to Chapters 4 and 5, and in particular to our discussion of Laplaces's principle of indifference and the choice of distinguishability versus indistinguishablity of individuals, we saw that this choice has to be resolved empirically. Indeed, we showed that a simple assembly rule for locating individuals in a spatially explicit landscape (see discussion of HEAP model in Box 4.2) generates the same results that would obtain if individuals were truly indistinguishable. Another assembly rule generates the outcomes expected for distinguishable objects.

How do other approaches to actually applying MaxEnt to macroecology compare to METE? Dewar and Porte (2008) use a constraint on resource consumption that is analogous to the one on energy that we use here. Their model differs from METE in that it requires input knowledge of the full distribution of resource consumption rates of the species. In contrast, METE predicts the distributions of metabolic rates across all individuals, $\Psi(\epsilon)$, across species, $\nu(\epsilon)$, and across individuals within species, $\Theta(\epsilon)$.

Dewar and Porte (2008), and also Pueyo et al. (2007) and Banavar and Maritan (2007), show that the logseries species–abundance distribution emerges from MaxEnt models if the appropriate prior knowledge is assumed. Pueyo et al. use prior probabilities to derive that result, whereas the other two papers make use of a related "maximum relative entropy" framework. In METE, the prediction of the logseries distribution is tightly linked to our choice of just two extensive (i.e. additive across systems) state variables: N_0 and E_0. The single power of n that appears in the denominator of the logseries distribution, $\Phi(n) \sim \exp(-\beta n)/n$, results from the integral over ϵ in Eq. 7.5. If we had introduced at the outset another conserved and limiting resource in addition to energy, say W for water, then we would have obtained $R(n, \epsilon, w) \sim \exp(-\lambda_1 n - \lambda_2 n\epsilon - \lambda_3 nw)$ and the integrals over ϵ and w would result in $\Phi(n) \sim \exp(-\beta n)/n^2$ (Exercise 10.1). Every additional extensive variable introduced into the theory to account for another limiting resource will result in an extra power of n in the denominator of $\Phi(n)$. That in turn will lead to a larger fraction of species predicted to have a single individual or fewer than some small number of individuals. Restating this, more relative rarity is predicted in environments with more resources that are limiting or constraining growth.

Thus, while other MaxEnt models predict the logseries' abundance distribution, the reasons they do are different from the reason METE does. Moreover, METE links energetics, diversity, abundance, and spatial scaling within a unified theoretical framework to predict relatively accurately, and without adjustable parameters, a much wider range of macroecological metrics than do the MaxEnt models referred to above.

Another application of MaxEnt to predicting the species–abundance distribution was proposed by Shipley et al. (2006). The premise and underlying strategy behind their work is straightforward. Consider a collection of species that possess some set of traits. For plants, for example, these traits might include leaf thickness, rooting depth, and other properties that influence plant fitness. Each species has associated with it a set of numbers giving the values of each of these traits. Hence we can form the average values of each of these traits by taking the abundance-weighted average of each trait value over the collection of species. In the context of MaxEnt, these abundance-weighted average trait values constitute the constraints. Shipley et al. then apply these constraints and derive from MaxEnt the species abundance distribution. Provided there are fewer traits than species, the problem is under-determined in a good sense: there will not be enough algebraic equations to uniquely determine each species' abundance. MaxEnt provides the additional information needed to derive the distribution of abundances.

This approach is useful in the sense that it can reveal the internal consistency of a MaxEnt application if the species–abundance distribution that results from the calculation is consistent with the abundance data used to generate the abundance-weighted trait averages. Given that the information contained in the prediction is already contained in the data needed to make the predictions, its practical value is less clear. A similar perspective pertains to a recent application of MaxEnt by Azaele et al. (2009).

10.5 Exercise

*Exercise 10.1

Suppose that METE is amended by the introduction of an additional state variable, W_0, which is just like E_0 in the sense that each individual has a value of w and the sum of the w-values over all individuals is W_0. Show that $\Phi(n)$ is now no longer the logseries distribution but rather depends on n as $\Phi(n) \sim e^{-\beta n}/n^2$.

11

Future directions

METE makes numerous predictions about patterns in macroecology, but it also leaves us with many as yet unanswered questions. I focus on three of these here: What can we infer about spatial correlations of individuals within ecosystems? What can we infer about ecological trophic network structure? How do macroecological patterns change over time?

11.1 Incorporating spatial correlations into METE

As formulated in Chapter 7, METE predicts only one aspect of the spatial distributions of individuals within species at multiple spatial scales. In particular, the metric $\Pi(n|A, n_0, A_0)$ describes the distribution of abundances, n, within a single cell of area A. This metric provides no information about joint distributions, such as the probability of simultaneously finding n_k individuals in cell A_k, where k can take on values of $1, 2, \ldots, K$ and the K cells are located at specified places within a landscape. The simplest metric of the latter type, and one often studied in ecology, is the "distance-decay" function, referred to in Section 3.3.3: it describes the dependence on inter-cell distance of the probability that a species is present in two cells of specified area. Other related measures of correlation are Ripley's K and the O-ring metric also discussed in Section 3.3.3. The values of all these metrics will, of course, be contingent on the abundance of the species at some landscape scale, the area of that landscape. The distance-decay metric will also depend on the area of the specified cells, while Ripley's K and the O-ring metric will depend on the radius of the circle surrounding a specified individual. Related to distance decay is the community-level commonality metric, which describes the fraction of those species found in a typical cell of area A that are expected to be in two such cells at distance D apart. As with distance decay, it will, in general, depend, at the very least, on the areas of the cells and the distance between them.

I suggest two possible approaches to calculating correlations within METE.

11.1.1 Method I: Correlations from consistency constraints

Here I focus on using METE to constrain the probabilities for every possible combination of occupancies in all the cells into which a plot has been subdivided. In other words, we are interested in inferring the metric $C(n_1, n_2, \ldots, n_k)$, which was mentioned in Section 3.3.3. It is the probability of observing a set of assigned abundances $\{n_1\}$ in all of the K cells that result when A_0 is gridded into cells of area $A = A_0/K$. I show that that knowledge of $\Pi(n)$ at more than one spatial scale imposes *consistency constraints* on the joint probabilities for finding different numbers of individuals simultaneously in more than 1 cell.

Consider the case of an area A_0, containing a species with $n_0 = 2$, and divided into four quadrants. This simple case is instructive because it is exactly solvable; knowledge of $\Pi(n)$ at scales $A_0/2$ and $A_0/4$ uniquely determine the probabilities of all possible assignments of the 2 individuals into the four cells. We show how this comes about first using the HEAP model (or equivalently the recursive METE predictions; see discussion following Eq. 7.51) for the $\Pi(n)$ and then using the non-recursive METE predictions. The procedure illustrated below is applicable, however, to any model or theory that predicts the $\Pi(n)$ metric.

Our starting point is the ten occupancy possibilities shown following Eq. 4.9 in Chapter 4. In HEAP, the probability of the first occupancy possibility, with 2 individuals in the upper-left quadrant, is given by 1/9 (see Box 4.2 and then do Exercise 11.1). Moreover, the probability of both individuals being in the left-half of the plot (the sum of the probabilities of occupancy possibilities 1, 2, and 5) is given in HEAP by 1/3 (the Laplace result). Therefore, because there are two occupancy possibilities with both individuals in a left-side quadrant, we have the consistency relationship:

$$2P\left(\begin{array}{|c|c|}\hline 2 & 0 \\ \hline 0 & 0 \\ \hline \end{array}\right) + P\left(\begin{array}{|c|c|}\hline 1 & 0 \\ \hline 1 & 0 \\ \hline \end{array}\right) = 1/3. \tag{11.1}$$

This implies that: $P\left(\begin{array}{|c|c|}\hline 1 & 0 \\ \hline 1 & 0 \\ \hline \end{array}\right) = 1/3 - 2/9 = 1/9$. We also have the normalization relationship:

$$4P\left(\begin{array}{|c|c|}\hline 2 & 0 \\ \hline 0 & 0 \\ \hline \end{array}\right) + 4P\left(\begin{array}{|c|c|}\hline 1 & 0 \\ \hline 1 & 0 \\ \hline \end{array}\right) + 2P\left(\begin{array}{|c|c|}\hline 1 & 0 \\ \hline 0 & 1 \\ \hline \end{array}\right) = 1, \tag{11.2}$$

implying that $P\left(\begin{array}{|c|c|}\hline 1 & 0 \\ \hline 0 & 1 \\ \hline \end{array}\right) = (1 - 4/9 - 4/9)/2 = 1/18$.

Running through the same calculations with the non-recursive METE values for $\Pi(n)$, for the diagram with both individuals in the upper-left quadrant we have $\Pi(n|A_0/4, 2, A_0) = 0.1161$ (Exercise 11.2). As with HEAP, the probability of both individuals being in the left-half of the plot (the sum of the probabilities of diagrams 1, 2, and 5) is given in METE by 1/3. Therefore, we have the consistency relationship:

$$2P\left(\begin{array}{|c|c|}\hline 2 & 0 \\ \hline 0 & 0 \\ \hline \end{array}\right) + P\left(\begin{array}{|c|c|}\hline 1 & 0 \\ \hline 1 & 0 \\ \hline \end{array}\right) = 1/3. \tag{11.3}$$

This implies that:

$$P\left(\begin{array}{|c|c|}\hline 1 & 0 \\\hline 1 & 0 \\\hline\end{array}\right) = 1/3 - 2.(0.1161) = 0.1011. \tag{11.4}$$

From the normalization relationship:

$$4P\left(\begin{array}{|c|c|}\hline 2 & 0 \\\hline 0 & 0 \\\hline\end{array}\right) + 4P\left(\begin{array}{|c|c|}\hline 1 & 0 \\\hline 1 & 0 \\\hline\end{array}\right) + 2P\left(\begin{array}{|c|c|}\hline 1 & 0 \\\hline 0 & 1 \\\hline\end{array}\right) = 1, \tag{11.5}$$

we derive

$$P\left(\begin{array}{|c|c|}\hline 1 & 0 \\\hline 0 & 1 \\\hline\end{array}\right) = [1 - 4.(0.1161) - 4.1011)]/2 = 0.0656. \tag{11.6}$$

In these examples, based on there being two individuals placed into 4 cells, there are just enough consistency constraints to determine the probabilities of all the distinct occupancy possibilities. More generally, at finer scales and with $n_0 > 2$, there will be both more possible occupancy possibilities (i.e. assignments of individuals) and there will be more constraint equations because they arise at every scale coarser than the one being examined. Thus the number of unknowns (probabilities of particular occupancy possibilities) will be greater and so will the number of equations

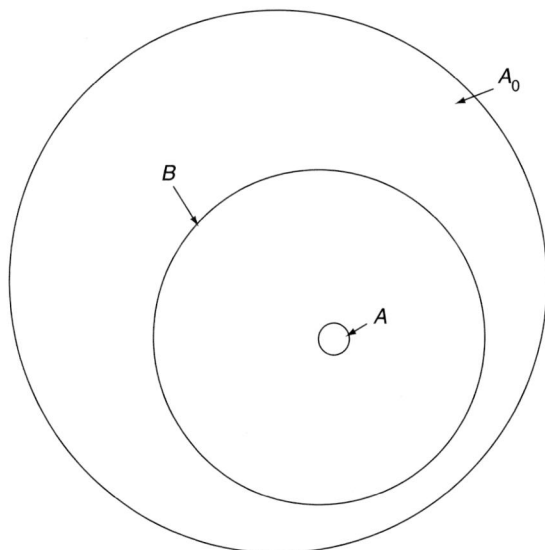

Figure 11.1 The geometry assumed in the derivation of Ripley's K and the O-ring metrics in METE. The calculation assumes that a single individual of a species is found in A, and answers the question: what is the probability that there are $n + 1$ individuals in the disk $\hat{B}$ = the torus B + the disk A.

determining those unknowns. Unfortunately, it can be shown that, in general, the number of unknowns exceeds the number of constraints when $n_0 > 2$ or $A > A_0/4$. Hence the probablity associated with each occupancy possibility cannot be uniquely calculated.

Nevertheless, these constraint equations must be satisfied and thus they provide relationships among occupancy possibilities with individuals spaced at different distances from each other. Conceivably, these constraints could be incorporated into a MaxEnt calculation and used to infer the most likely distance-decay function. Doing so will require much more effort than is expected for a homework exercise; perhaps an inspired reader will figure out how to complete this agenda!

11.1.2 Method 2: A Bayesian approach to correlations

Here I sketch a method for calculating Ripley's K and the O-ring metric within METE based on Bayes' Law. I start with an estimate for the species-level versions of these metrics and then extend the result to community level.

Consider a small cell or area A nested at the center of a much larger cell of area B, which in turn is nested within an even larger area A_0, for which the state variables S_0 and N_0 are known. For simplicity we take these cells to be circles (Figure 11.1). We focus on an arbitrary species that has abundance n_0 in A_0. To make the derivation below as simple as possible, we assume that area A is small enough that it can contain only 0 or 1 individuals of the species. This assumption can be relaxed and a useful answer can still be derived (Exercise 11.4). We use the notation B to indicate both the region B and the area of that region. Importantly, we take the region B to be a two-dimensional torus, not a disk, and use the notation $\hat{B}$ to refer to the equivalent disk with area A included. Hence $\hat{B}=B + A$.

Consider the joint probability that there is 1 individual of the species in A and $n + 1$ individual of that same species in $\hat{B}$ (one of which must be the one in A): $P(n+1$ in $\hat{B}$, 1 in A). Bayes' Law says that we can write this in two equivalent ways:

$$P(n+1 \text{ in } \hat{B}, 1 \text{ in } A) = P(n+1 \text{ in } \hat{B}|1 \text{ in } A) \cdot P(1 \text{ in } A) \\ = P(1 \text{ in } A|n+1 \text{ in } \hat{B}) \cdot P(n+1 \text{ in } \hat{B}). \quad (11.7)$$

Here, for notational convenience, we have left out the conditionality condition that there are n_0 individuals in A_0, but it is implicit.

From the definition of the metric Π, we can write (with the conditionality on n_0 now explicit on the right-hand sides of Eqs 11.8, 11.9):

$$P(n+1 \text{ in } \hat{B}) = \Pi(n+1|\hat{B}, n_0, A_0), \quad (11.8)$$

and

$$P(1 \text{ in } A) = \Pi(1|A, n_0, A_0), \quad (11.9)$$

$$P(1 \text{ in } A | n + 1 \text{ in } \hat{B}) = \Pi(1|A, n+1, \hat{B}). \tag{11.10}$$

The right-hand sides of these three equations are all predicted by METE. Noting that $P(n \text{ in } B | 1 \text{ in } A) = P(n + 1 \text{ in } \hat{B} | 1 \text{ in } A)$, Eqs 11.8–11.10 now imply:

$$P(n \text{ in } B | 1 \text{ in } A) = \frac{\Pi(1|A, n+1, \hat{B}) \cdot \Pi(n+1|\hat{B}, n_0, A_0)}{\Pi(1|A, n_0, A_0)}. \tag{11.11}$$

If cell A is still nested within cell B but large enough to contain more than 1 individual of any particular species, we can write a revised version of Eq. 11.11. It will be replaced with an expression for $P(n \text{ in } B | m \text{ in } A)$ (Exercise 11.4).

Next, we calculate the expected number of individuals in B if the species is found in A, which we denote by $\bar{n}_{B/A}$:

$$\bar{n}_{B/A} = \sum_{n=0}^{n_0-1} n \cdot \frac{\Pi(1|A, n+1, \hat{B}) \cdot \Pi(n+1|\hat{B}, n_0, A_0)}{\Pi(1|A, n_0, A_0)}. \tag{11.12}$$

To calculate the average number of individuals in B subject to the species being found in A when there is no limit, other than that imposed by n_0, on the abundance that can be found in A (i.e. the revised Eq. 11.12), we now have to perform a double summation over the abundance, m, in A, where $m \geq 1$, and the abundance n in A. The former sum will extend from 1 to $n_0 - n$ and the latter sum extends from 0 to $n_0 - 1$.

If Eq. 11.12 is evaluated using the METE result for the Πs, the dependence of expected abundance in B on three variables: n_0, $A_0/\hat{B}$, and $\hat{B}/A$ will result. The dependence of Ripley's K (see Section 3.3.3) on these quantities is thus determined. To obtain the O-ring metric, it is simply necessary to evaluate this expression for two different-sized, torus-shaped cells, B and C, and subtract the smaller from the larger to get the expected number of individuals in the O-ring formed between two tori.

This Bayesian approach can be extended to predict the expected number of species that are found in both B and A, $S(A \cap B)$. If we assume that there is no limit on the number of individuals that A contains, and follow the same logic that led to Eq. 11.12, we can calculate $S(A \cap B)$ as a double sum over m and n_0 of $\Phi(n_0)[1 - P(0 \text{ in } B | m \text{ in } A)]$, where $\Phi(n_0)$ is the species–abundance distribution in A_0 and $P(0 \text{ in } B | m \text{ in } A)$ is calculated as described in the previous paragraph:

$$S(A \cap B) = S_0 \sum_{n_0=1}^{N_0} \sum_{m=1}^{n_0} \left\{ 1 - \frac{\Pi(m|A, m, \hat{B}) \cdot \Pi(m|\hat{B}, n_0, A_0)}{\Pi(m|A, n_0, A_0)} \right\} \cdot \Phi(n_0). \tag{11.13}$$

A related expression can be derived for the expected number of species found in both A and an O-ring between two cells B and C (Exercise 11.6).

11.2 Understanding the structure of food webs

As pointed out in Section 10.3, the derivation of the linkage distribution, $\Lambda(l|S_0, L_0)$, from MaxEnt uses as state variables the total number of linkages in the web and the total number of nodes. The number of nodes is just our state variable, S_0, while the total number of linkages, L_0, is an additional state variable. Is it possible, however, that METE could be used to predict the state variable L_0?

And here is another unanswered question about food webs. Link distributions give us information about the topological structure of food webs, but they include no information about the distribution of the magnitudes of energy flow in the trophic links. Can we use METE to predict not just the qualitative or topological structure of webs, but also the distribution of the strengths of the linkages?

Consider a bipartite web, such as that linking a community of plants to a community of herbivores, and assume that at each of those two trophic levels, the number of species, the number of individuals, and the total metabolic rate are specified. Call them $S_{0,P}$, $N_{0,P}$, $E_{0,P}$ for the plants and $S_{0,H}$, $N_{0,H}$, $E_{0,H}$ for the herbivores. From these state variables, METE predicts the abundance distribution and the distribution of metabolic rates across species at each of the levels. For all the individual herbivores to meet their metabolic requirements, $E_{0,H}$ in total, they have to eat a sufficient quantity of the product of plant metabolic use (i.e. growth). To achieve that, herbivores will require sufficient linkages to plants to ensure that the metabolic needs of all the herbivores are met. If a value of herbivore food-assimilation efficiency (metabolic energy derived/energy of food eaten) is assumed, then the distribution of flow rates across the network will be constrained by the fact that the average flow rate to the herbivores is proportional to $E_{0,H}/L_0$. MaxEnt will then predict an exponential distribution of flow rates from plant nodes to herbivore nodes.

Now the problem boils down to determining what L_0 needs to be so that when plant outputs are matched with needed herbivore inputs, the energy requirements of the herbivores are met. Doing so appears to require the adoption of some rule that assigns the way in which that matching occurs. In the spirit of neutrality, one such rule could be that all trophic choices are random, without regard to whether large mass herbivores eat small plants or large plants. Another possible rule might be that largest eats largest. With whatever rule is chosen, trophic needs can be sequentially met until all needs are met and at that point, the number of links it took to meet those metabolic needs is the predicted L_0.

In principle, this strategy could be extended from bipartite webs to full trophic webs, although this is clearly not a simple problem.

11.3 Toward a dynamic METE

> *In the knowledge derived from experience.*
> *The knowledge imposes a pattern, and falsifies,*
> *For the pattern is new in every moment . . .*
> **T. S. Eliot,** "East Coker," from the *The Four Quartets*

Imagine trying to apply the ideal gas law, $PV = nRT$, in the atmosphere during a tornado! The thermodynamic state variables would all be changing rapidly; the system, however defined, would be out of equilibrium. Exaggerating somewhat, ecosystems are like tornadoes, ever changing, never in equilibrium. So how can we justify METE, a theory based on static values of state variables? Strictly speaking we really can't. In fact, as we noted in Chapter 8, some of the failures of METE may well be a consequence of the censused systems being sufficiently not in steady state, with the state variables changing sufficiently rapidly, so that predicting spatial patterns or abundance and metabolic rate distributions is like applying $PV = nRT$ in a tornado.

To know whether that is the case, and to extend the applicability of the theory if it is the case, requires development of a far-from steady state version of METE. We can think of METE, as described up to this point, as being a theory of macroecological patterns at slices in time, with the patterns determined by the values of the state variables at those moments in time. If the state variables change sufficiently slowly, it is plausible to imagine that the changing state variables slowly drag the patterns with them and the predictions are valid. But if the state variables are changing sufficiently rapidly, then the patterns may not be predicted from instantaneous values of the state variables.

So the first step in making METE dynamic is development of a theory that predicts how state variables change in time. Several approaches to developing a dynamic theory of state variables could be explored. We might turn to conventional models of ecosystem dynamics, from Hubbell's neutral theory (Hubbell, 2001) to more mechanism-rich models that could potentially describe how S_0, N_0, and E_0 change over time. There are serious limitations in every such approach we know of. For example, the neutral theory assumes a zero sum constraint, implying that N is by definition static. That assumption could be relaxed but the theory we seek is precisely what is needed to tell us how to relax it. Second, speciation rate in the theory is governed by an assumption and determined by fitting data; it is not really predicted from anything more fundamental. Third, the theory in its present form does not include a description of growth in the sizes of individuals within or across species nor of temporal change in metabolic rates, and therefore it does not readily provide information about the dynamics of our state variable E_0.

An intriguing, but unabashedly speculative, approach involves the concept of maximum entropy production (MEP). The basic idea of MEP, and the uncertainty surrounding its validity, were described briefly in Section 6.3.5. MEP states that a physical system far from equilibrium, undergoing an irreversible transition from one macroscopic state to another, will most likely choose a path in phase space that maximizes the rate of production of entropy. Essentially, what this means is that physical systems, far from equilibrium, tend to dissipate as much energy in the form of heat as they possibly can. So, colloquially speaking, a stream cascading down a hillside will take a path that scours the hillside as much as possible. Wind blowing heat from Earth's sun-heated equator to the poles will do so at a rate that maximizes

Box 11.1 Simple version of the Paltridge model of latitudinal heat convection and MEP

Figure 11.2 sets up the model. Were it not for convective flow of warm air and surface water from the equator to the poles, the equator would be much hotter than it now is and the poles much cooler. Let Q be the rate of heat flow from the equator to each pole. T_e and T_p label the temperature of the equator and the pole in the presence of convective heat readjustment. The net solar radiation fluxes to the equatorial and polar regions are labeled Ω_e and Ω_p. By net flux we mean the difference between the incoming flux and the reflected outgoing flux, with flux measured in units of, say, watts per square meter. We denote entropy by the symbol S, risking confusion with our symbol for species richness in order to maintain agreement with thermodynamic convention. Then, using the time derivative of Eq. 5.1, the rate of production of entropy, dS/dt, as a consequence of the heat flow to the far northern latitudes is the difference between the rate of entropy change at the equator, dS/dt $= -Q/T_e$, which arises because of the flow of heat away from the equator, and its rate of change when that heat is deposited at high latitude: dS/dt $= Q/T_p$. Note that because $T_p \leq T_e$, the net rate of entropy production, dS/dt:

$$\frac{dS}{dt} = \frac{Q}{T_p} - \frac{Q}{T_e} \tag{11.14}$$

is ≥ 0. Entropy is produced when heat is added to an object, explaining the + sign in front of the term $\Delta Q/T_p$. A term similar to Eq. 11.14 would describe the rate of entropy production in the southern hemisphere.

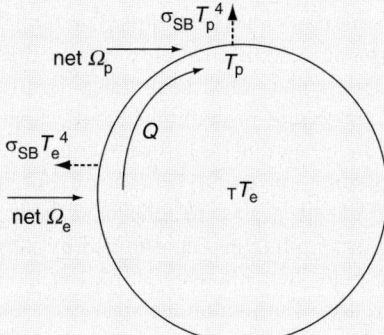

Figure 11.2 Planetary heat convection from equator to pole. Ω_e and Ω_p are the net solar radiation fluxes at the equator and pole respectively, in units of watts/m². The net flux is the incoming solar flux – the outgoing reflected solar radiation. The terms $\sigma_{SB}T_e^4$ and $\sigma_{SB}T_p^4$ are the outgoing infrared radiation fluxes, where σ_{SB} is the Stefan Boltzmann constant. Q is the convective heat flux from equator to north polar region; a similar convective heat flow and radiation flux are not shown for the south polar region.

Our goal is to find the value of Q that maximizes the expression in Eq. 11.14. To do so, we have to re-express T_e and T_p in terms of Q. Toward that end, we introduce the equations of energy conservation for the equatorial and polar regions:

$$\Omega_e = 2Q + \sigma_{SB}T_e^4, \tag{11.15}$$

Box 11.1 (*Cont.*)

$$\Omega_p + Q = \sigma_{SB} T_p^4. \tag{11.16}$$

The quantity σ_{SB} is the Stefan–Boltzmann constant, and the factor of 2 in front of Q in Eq. 11.15 arises because a flow of heat equal to Q exits the equatorial region in each of the two poleward directions.

Using Eqs 11.15 and 11.16, we can eliminate both T_e and T_p in Eq. 11.14, and arrive at an equation for the rate of entropy production, dS/dt, solely in terms of the rate of heat flow, Q. We can then maximize dS/dt by setting to zero its derivative with respect to Q, and of course checking that the result is a maximum, not a minimum. We leave the details of the rest of the calculation to the reader (Exercise 11.9).

The result of the calculation is shown in Figure 11.3. The value of Q obtained from this calculation is remarkably close to the measured average value. If the calculation is done within a more detailed energy balance model, in which the hemispheres are divided in to latitudinal bands, the accuracy improves and the latitudinal dependence of Q is fairly accurately predicted. Moreover, substitution of the values of Q obtained from the calculation into Eqs 11.15, 11.16 yields values of T_e and T_p that are in reasonable agreement with measured average values.

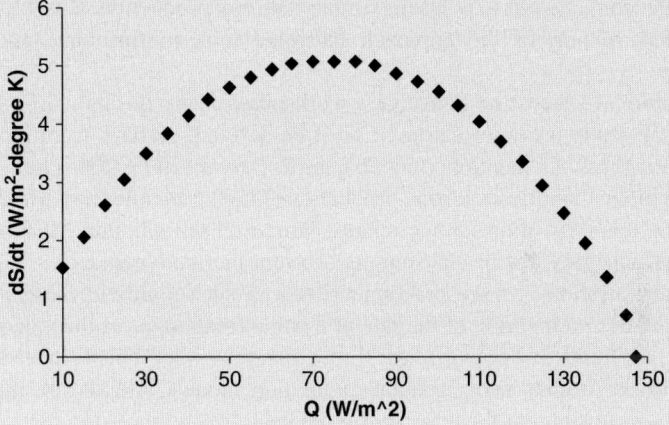

Figure 11.3 The rate of entropy production, dS/dt, versus the rate of poleward heat flow, Q. The parameters were chosen as: $\Omega_p = 50$ W/m², $\Omega_e = 500$ W/m². Entropy production is maximized at a poleward heat flow of $Q_{MEP} = 75$ W/m², which gives $T_p = 217$ K, $T_e = 290$ K.

Interestingly, if the values of the net incoming solar fluxes are varied and different values of the Q-value that maximizes entropy production are plotted against the magnitude of the resulting temperature gradient $(T_e - T_p)$ that is obtained at MEP, the relationship is not linear; in other words, far from equilibrium, the flows predicted by imposing the constraint of MEP are not linearly proportional to the strength of the gradients that produce these flows (Exercise 11.10).

the production of disorder as quantified by the Carnot definition of entropy, $\Delta Q/T$ (Eq. 5.1), associated with sensible heat transfer. Here we illustrate and pursue further this notion of MEP.

Imagine the flow of a fluid driven by the temperature difference between two locations: the hotter location is heated by some fixed heat source and the other, cooler one, is not. The fluid flowing from the hotter location to the colder one will convey heat, tending to cool the warm location and warm the cool one. If the flow is too great, the temperature gradient will disappear and then there will be no more flow. So, too large a flow is not sustainable. If the flow is too weak, the temperature difference will be large, and then more flow will ensue. Could there be some law of nature that determines where, between these two extremes of too low or too high a convective flow, nature tends to reside?

If the heat flow were not conveyed by a convective, possibly turbulent, fluid, but rather was conducted in, say, a solid, then the problem is simpler. Fick's law states that the rate of heat flow will be proportional to the temperature gradient and the rate of heat flow can be readily calculated. But if a moving and turbulent fluid conveys the heat, then the problem is much messier.

Many years ago, Garth Paltridge (1975, 1978, 1979) proposed that the convective flow of heat from equator to poles maximizes entropy production. Box 11.1 presents a simplified version of the approach Paltridge took to formulate and test his hypothesis.

The empirical success of Paltridge's work raises many questions. First, we can ask: Is MEP really a law of nature? Can it be derived, perhaps from more fundamental principles? As mentioned in Chapter 5, Dewar (2003, 2005) has outlined a derivation of the maximum entropy production (MEP) principle from MaxEnt. As of this writing, his derivation is incomplete. Time will tell whether MEP attains the status of an accepted law of far-from-equilibrium thermodynamics.

A further question is more practical: How can such a simple calculation work when it includes no mention of the rate of Earth's rotation, or of the viscosity of air and water, the fluids that convect heat from equator to pole? This is indeed a puzzle. Perhaps further studies using general circulation models (GCMs) of atmospheric dynamics will reveal that the equator-to-pole heat flow predicted in these models is relatively insensitive to rotation rate and viscosity. It is unimaginable that the flow is completely independent of rotation rate and viscosity, but if the dependence is relatively weak that might explain why the MEP calculation "works." On the other hand, the success of MEP could be accidental. But in its favor, MEP is supported by considerable data from many sources (see, for example, Kleidon and Lorenz, 2005) and not just from the Paltridge calculation.

For the remainder of this Chapter, I shall assume that the Dewar proof will ultimately be successful and that MEP is a law of far-from-equilibrium thermodynamics. Several authors have begun to explore the implications of this possibility. I refer readers to the collection of chapters in the edited collection by Kleidon and Lorenz (2005), and also to Kleidon et al. (2010), for descriptions of possible applications to Earth system science. The application I pursue here is quite different, however, from current work in the field. It begins with the observation that Dewar's work suggests that MEP, if valid, should apply to the rate of production of information entropy, as well as to thermodynamic entropy. What, then, are the implications of assuming that the production of information entropy far from steady state is governed by an extremum principle? Let us explore this.

We start with the fundamental probability distribution of METE, $R(n, \epsilon)$, which from Chapter 7 is given by:

$$R(n, \varepsilon | S_0, N_0, E_0) = \frac{e^{-\lambda_1 n} e^{-\lambda_2 n \varepsilon}}{\sum_{n'=1}^{N_0} \int_0^{E_0} e^{-\lambda_1 n'} e^{-\lambda_2 n' \varepsilon'} d\varepsilon'}. \quad (11.17)$$

The Lagrange multipliers, λ_1 and λ_2, contain the dependence on the state variables, S_0, N_0, and E_0.

The information entropy associated with this distribution is:

$$S_I = -\sum_n \int d\varepsilon R(n, \varepsilon) \log(R(n, \varepsilon)). \quad (11.18)$$

Recalling results from Chapter 7, provided $S_0 \gg 1$, the quantity $\beta = \lambda_1 + \lambda_2$ is determined from:

$$\beta \log(\frac{1}{\beta}) \approx \frac{S_0}{N_0}. \quad (11.19)$$

The second Lagrange multiplier is given by:

$$\lambda_2 = \frac{S_0}{E_0 - N_0}. \quad (11.20)$$

To simplify the notation in what follows, we leave the subscript 0 off the state variables and we define:

$$\omega \equiv \log\left(\frac{1}{\beta}\right). \quad (11.21)$$

Then, carrying out the integration and summation in Eq. 11.18, it can be shown (Exercise 11.11) that:

$$S_I = 1 + \log\left(\frac{E}{N}\right) + \omega + \omega^{-1}. \quad (11.22)$$

We now explore the implications of assuming that the time derivative of information entropy (that is, the rate of entropy production) is an extremum. This, we will see, leads to equations that, in principle, determine the rates of changes of the state variables S, N, and E as functions of those same variables. Solutions to these equations could then provide a basis for a dynamic theory of macroecology.

The time derivative of S_I is given by (Exercise 11.12):

$$\frac{dS_I}{dt} = \omega^{-1}\frac{1}{N}\frac{dN}{dt} + \frac{1}{E}\frac{dE}{dt} - (1+\omega^{-1})\frac{1}{S}\frac{dS}{dt}. \tag{11.23}$$

To derive the consequences of our assumption, we find the extremum of this expression wth repect to variation in the state variables. Setting to zero the derivatives of dS_I/dt with respect to each of the three state variables, we then obtain three coupled partial differential equations for the partial derivatives, with respect to state variables, of the time derivatives of the state variables. These partial derivatives are expressed as functions of the state variables and their time derivatives. The solutions to these equations will then be a set of coupled differential equations relating the time derivatives of the state variables to the state variables themselves. The solutions to those coupled equations describe the time evolution of the state variables under the assumption of MEP.

Here are the equations that result from setting each of the three derivatives of dS_I/dt equal to zero:

$$(1+\omega)\frac{\partial(dS/dt)}{\partial S} = \left(\frac{1}{\omega-1}\right)\frac{dN/dt}{N} + \omega\frac{S}{E}\frac{\partial(dE/dt)}{\partial S} + \frac{S}{N}\frac{\partial(dN/dt)}{\partial S} + \left(\frac{\omega^2-2}{\omega-1}\right)\frac{dS/dt}{S}, \tag{11.24}$$

$$\left(\frac{\omega^2-1}{\omega}\right)\frac{N\partial(dS/dt)}{S\ \partial N} + \frac{dN/dt}{N} = (\omega-1)\left(\frac{N}{E}\right)\frac{\partial(dE/dt)}{\partial N} + \left(\frac{\omega-1}{\omega}\right)\frac{\partial(dN/dt)}{\partial N} + \omega^{-1}\frac{dS/dt}{S}, \tag{11.25}$$

$$(1+\omega)\frac{E}{S}\frac{\partial(dS/dt)}{\partial E} + \omega\frac{dE/dt}{E} = \omega\frac{\partial(dE/dt)}{\partial E} + \frac{E}{N}\frac{\partial(dN/dt)}{\partial E}. \tag{11.26}$$

The solutions to these equations yield an extremum for dS_I/dt, but that extremum must be a constant in time. The reason is that time dependence in dS_I/dt can only arise from time dependence of the state variables. But the equations enforce the independence of dS_I/dt on those state variables.

Consider now the special case in which dN/dt and $dE/dt = 0$. It then follows from Eq. 11.23, and the fact that the extremum of dS_I/dt is some constant, C, that the equations reduce to the simpler equation:

$$(1+\omega^{-1})\frac{1}{S}\frac{dS}{dt} = -C \tag{11.27}$$

You can show (Exercise 11.14) that for $C > (<) \, 0$, this equation results in $dS/dt < (>) \, 0$, and thus declining (increasing) species richness. Hence, for the case of constant N and E, and the rate of change of information entropy given by a constant, species diversification corresponds to decreasing information entropy. The MEP solution, with C as large as possible, corresponds to species richness declining as fast as possible!

If N and E are allowed to vary in time, then more complicated dynamics will arise. The full implications of Eqs 11.24–11.26 have yet to be explored. Another issue raised, but not answered, by our imposition of MEP on the information entropy function $R(n,\epsilon)$ is the intrinsic time scale for change. Our equations have a parameter, t, for time, but because there is no other quantity in the theory with dimensions of time, it is not clear what the absolute time scale of change is. Because E is a rate of metabolism, the ratio M/E ratio, where M is the biomass of the system (recall mass$_{min} = \epsilon_{min} = 1$), might set an intrinsic time-scale.

Clearly we have speculated here well beyond available evidence. MEP is as yet unproven. If it is true, its applicability to information entropy is uncertain. If it does apply to information entropy, it is not clear why we should apply it to $R(n,\epsilon)$, as we did above, rather than to, say $\Pi(n)$ or $\Phi(n)$ or $\psi(\epsilon)$. Or perhaps it should be applied to all these probability distributions. It is also unclear, even if nature selects an extremum of entropy production, whether that extremum is a minimum or a maximum. And finally, we have not explored all the implications of Eqs 11.24–11.26, nor tested any of them.

Yet, there is some appeal to the notion that MEP might provide the basis for a dynamic extension of METE. In particular, at the core of Dewar's partial derivation is the MaxEnt principle. MaxEnt is applied to infer the probability distribution, defined over final macrostates, of the number of paths in phase space leading irreversibly from an initial macrostate to each of the possible final macrostates. If more paths lead to the final macrostate that corresponds to MEP, then that final state is most likely. Substituting "paths in phase space" for "number of microstates in phase space" reveals the logical connection between MEP and the classical statement of the second law of thermodynamics.

11.4 Exercises

Exercise 11.1

Show that in HEAP, $\Pi(2|A_0/4, 2, A_0) = 1/9$ and that in non-recursive METE, $\Pi(2|A_0/4, 2, A_0) = 0.1161$.

Exercise II.2

Again, for the case in which $A = A_0/4$ and $n_0 = 2$, work out the implications of all the cross-scale consistency constraints for the random placement model. Explain why you could have written down your results for $\begin{pmatrix} 1 & 0 \\ 1 & 0 \end{pmatrix}$ and $\begin{pmatrix} 1 & 0 \\ 0 & 1 \end{pmatrix}$ right away, without looking at the consistency constraints.

Exercise II.3

Show that when A_0 is subdivided into quadrants, then for any $n_0 > 2$ the number of possible diagrams assigning individuals to cells exceeds the number of cross-scale constraints and thus the problem is underdetermined (try the simplest case in which $n_0 = 3$ to get you started). Then show that when $n_0 = 2$, and A_0 is subdivided into 2^j cells with $j > 2$, the problem is again underdetermined (try the simplest case in which $j = 3$).

Exercise II.4

Carry out the calculation described following Eq.11.11 to derive $P(n$ in $B|m$ in $A)$ if m is limited only by n_0.

Exercise II.5

Derive the revised Eq. 11.12 for the case in which only n_0 limits the number, m, of individuals in A.

Exercise II.6

Derive Eq. 11.13.

Exercise II.7

Derive an expression for the O-ring metric following the approach described after Eq. 11.12. Do this for the case in which only n_0 limits the possible value of m.

**Exercise 11.8

Substitute into your answer for Exercise 11.7 the METE solution to $\Pi(n)$ and determine the dependence of the O-ring metric on the radius of the ring. How does the form of that distance dependence itself depend upon the abundance, n_0, of the species?

Exercise 11.9

Find the function $Q(\Omega_e, \Omega_p)$ that maximizes the rate of entropy production in the Paltridge model.

Exercise 11.10

Using your result in Exercise 11.10, plot the value of Q_{max} as a function of the equator-to-pole temperature difference $(T_e - T_p)$. To derive different values for, Q_{max} T_e, and T_p, you will have to assume a variety of values for Ω_e and Ω_p.

Exercise 11.11

Derive Eq. 11.22.

Exercise 11.12

Derive Eq. 11.23.

*Exercise 11.13

Derive Eqs 11.24–11.26.

*Exercise 11.14

To explore numerically the solution to Eq. 11.27, it helps to make an analytical approximation to the solution to Eq. 11.19. The following is fairly accurate:

$\omega \approx \log(N/S) + \log(\log(N/S))*(1+1/\log(N/S))$, provided $N/S \gg 1$. Making this approximation, solve Eq. 11.27 numerically and discuss the dependence of the rate of change of S on the fixed value of N and on the constant, C.

*Exercise 11.15

What can be learned about the dependence of the time derivatives of the state variables on the state variables themselves using Eqs 11.24–11.26 if other combinations of the state variables than N and E are assumed to be fixed constants?

**Exercise 11.16

What can you learn from Eqs 11.24–11.26 if no constraints on the state variables are assumed?

**Exercise 11.17

Try to implement the strategy outlined in Section 11.2 to determine the number of linkages and the distribution of flow rates across them in a plant-herbivore web as a function of arbitrarily assigned state variables S, N, E at each of the two trophic levels.

Epilogue

Is a comprehensive unified theory of ecology possible? What might it look like?

If you stare at data you may see patterns, and if you think long and hard enough about patterns you may see explanations. That is how theory often arises in science. It is the (admittedly oversimplified) trajectory that took us from Tycho Brahe's notebooks full of raw data on the positions of the planets over time, to Kepler's ellipses and other regularities in the orbits, to Newton's law of gravity. The development of the theory of evolution also followed, at least loosely, that trajectory; remarkably, in that case major advances in data acquisition and pattern discernment, as well as theory building, were to a considerable extent all the work of one person!

Evolution is at the core of all of biology, and if a grand theory of ecology were to emerge, evolution would be an essential component. Is it possible that evolution is not only necessary but also sufficient to explain, and to be able to predict, everything that is predictable about ecology? What role, if any, might ideas discussed in this book play? So we return to some of the questions raised in Chapter 1, armed with what we have learned about patterns in macroecology.

First, let's review what we have accomplished here. METE does predict a lot with a little. Using only the logic of inference, we have developed a theory that predicts, without adjustable parameters, the central tendencies of the major patterns in the distribution, abundance, and energetics of individuals and species across wide ranges of spatial scale. It appears to work fairly well across taxonomic groups and across habitat types. It appears also to provide insight into inter-specific trophic network structure. In some cases, there are sizeable deviations from predicted patterns, but there appears to even be pattern in the deviations. So maybe we have the beginnings of a theory of macroecology. Far more testing is needed.

But METE is incomplete. The state variables, S, N, and E, which are central to the theory, are taken as given; they are not explained. What determines their values and, more importantly, what determines how each of the state variables changes in time? METE is a static theory of macroecology, not a dynamic theory of all of ecology.

Mechanism does not play a role in METE except insofar as it may be needed to determine the values of the state variables and their rates of change.

Section 11.3 advances the idea that the principle of maximum entropy production (MEP), applied to information entropy, could be the basis for a dynamic METE. But that would still not be a comprehensive, unified theory of ecology, for it would not describe, let alone explain, a vast amount of ecology: feeding and mating behavior, social organization, physiology and growth dynamics, adaptation, speciation, extinction, responses of organisms to environmental stress. Evolution must be a part of such a theory.

The Figure below is no more than a vision, and is appropriate only to an epilogue. It shows how I think ecology might develop in the coming decades. Evolution, without which "nothing makes sense in biology" (Dobzhansky, 1964), is the major player, but information theory in the form of MaxEnt and MEP plays an important role. MaxEnt, in combination with measured values of the state variables, predicts macroecological patterns, while MEP, perhaps derived from MaxEnt (see Sections 6.3.5 and 11.3), predicts the dynamics of the state variables. I have thrown in for good measure the possibility that MEP will provide a firmer basis than we have at present for understanding how metabolic rates depend on body mass. The explanations provided in this vision by MEP are shown by light dashed lines because today these are speculative.

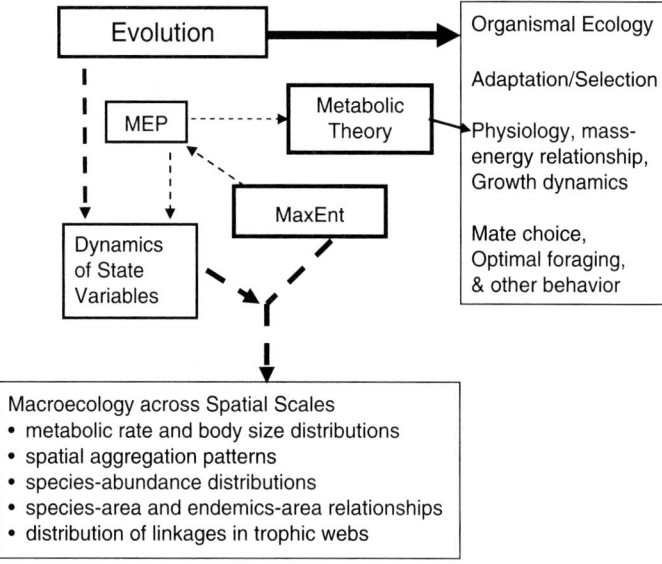

A model of a possible theory of ecology. Heavy dashed links are developed in this book; light dashed links are in progress or speculative; solid links are well established. The link from "MEP" to "Dynamics of State Variables" is discussed in Section 11.3.

As if the above is not sufficiently speculative, I will end even further out on a limb. Information theory appears to be infiltrating science. New developments suggest that information entropy may help us better understand everything from gravity to consciousness. Just possibly, Wheeler saw far in to the future when he suggested "its from bits." Perhaps decades from now all of the sciences, including ecology, will appear more similar and unified than they do today because of the ubiquitous role that information entropy will play in the explanation of patterns in all of nature. Of course there is no substitute in biology for the fundamental role of Darwin's grand idea, the inheritability of organismal traits that differ among individuals and that confer fitness. But the power of that idea is not diminished when it is augmented with the predictive capacity that flows from the logic of inference.

Appendix A

Access to plant census data from a serpentine grassland

The following website contains a spatially-explicit vegetation dataset:
   http://conium.org/~hartelab/MaxEnt.html

The census was carried out over the spring and summer of 1998 by Jessica Green on a 64 m^2 plot at the University of California's McLaughlin Reserve in northern Lake Napa County and Southern Lake County, CA. The plot was gridded to a smallest cell size of ¼ m^2 and in each cell the abundance of every plant species found there was recorded. The spreadsheet found at the website is organized as follows: the columns are plant species, with each species given a code name explained below the table of data. There are 256 rows of data, with each row corresponding to one of the ¼ m^2 cells. If the plot is viewed as a matrix, then the first row of data in the spread sheet corresponds to the upper-left cell (matrix element a_{11}). The second row of data is the matrix element a_{12}, or in other words the cell just to the right of a_{11}. The 17th row of data then corresponds to the plot matrix element a_{21}, and the very last row of data is the lower right cell, $a_{16,16}$. The actual data entries are the abundances of the species in each cell.

The data may be used by readers for any purpose, but any publication that includes use of the data should reference the dataset to:

   Green, J., Harte, J., and Ostling, A., (2003). Species richness, endemism, and abundance patterns: tests of two fractal models in a serpentine grassland. *Ecology Letters* 6, 919–928.

Moreover, the Acknowledgments should include a thanks to Jessica Green for use of the data.

Appendix B

A fractal model

In the four sections that follow, the formulation of a fractal (scale-invariant) model for many of the metrics of macroecology is presented.

B.1 A fractal model for $B(A|n_0, A_0)$

The method of incorporating scale invariance into macroecology explained below stems from pioneering work by W. Kunin (1998) that was further extended by Harte et al. (1999a, 2001). To define the model, we first introduce a notation for labeling cells obtained from A_0 by successive bisections. We label the cells by an index i, such that $A_i = 2^{-i} A_0$. We use the symbol A both to refer to a cell and to refer to the area of the cell. Each species is assigned a probability function $\alpha_j(n_0, A_0)$, where j is the same type of scale label as is i. α_j is the probability that if that species is present in A_0 with n_0 individuals, and also is found in a grid cell of area A_{i-1}, then it is present in a pre-specified one of the two A_i cells (say the one on the left) that comprise the A_{j-1} cell. Defining:

$$\lambda_i(n_0, A_0) \equiv \prod_{j=1}^{i} \alpha_j(n_0, A_0), \qquad (B.1.1)$$

it is easy to show (Exercise 4.7) that $\lambda_i(n_0, A_0) = 1 - \Pi(0|A_i, n_0, A_0)$, or in other words, λ_i is the probability that the species is present in an arbitrarily chosen A_i cell if the species has n_0 individuals in A_0 (Harte et al., 2001, 2005).

Combining this with Eq. 3.7, which relates $B(A)$ to $\Pi(0)$, the range–area relationship for each species is uniquely specified in terms of the λ-parameters:

$$B(A_i|n_0, A_0) \equiv A_0 \alpha_i(n_0, A_0). \qquad (B.1.2)$$

Scale-independence (or self-similarity) can now be imposed by assuming that the α_j are independent of scale, j, so that each $\alpha_j(n_0, A_0) \equiv \alpha(n_0, A_0)$ In words, scale-independence of the α_js means that the presence or absence of a species in a half-cell of an occupied cell is independent of cell area. In that case, the box-counting measure of occupancy, is related to cell area by:

$$B(A_i|n_0, A_0) \equiv A_0[\alpha(n_0, A_0)]^i. \quad (B.1.3)$$

Recalling that $A_i = A_0/2^i$, Eq. B.1.3 is mathematically equivalent (see Exercise 4.8) to the power-law form:

$$B(A_i|n_0, A_0) \sim A_i^y, \quad (B.1.4)$$

with

$$y = -\log_2(\alpha(n_0, A_0)). \quad (B.1.5)$$

The spatial distribution of each species then has a fractal dimension that is a function of the value of $\alpha(n_0, A_0)$ for that species. In particular, the fractal dimension of the spatial distribution, D, is related to y by $D = 2(1-y)$.

To determine these $\alpha(n_0, A_0)$ empirically, one would fit range–area data for each species to Eq. B.1.4; the slope of the $\log(B)$ vs. $\log(A_i)$ graphs will give an estimate of the $\alpha(n_0, A_0)$ using Eq. B.1.5.

The fractal model we have presented here was based on a specific method of partitioning space with successive bisections. Maddux (2004) criticized this partitioning because if interpreted too literally, it can lead to unrealistic patterns; Ostling et al. (2004) showed how the concerns raised by Maddux can be circumvented by taking predictions for a cell A_i to apply to an average cell of that area placed anywhere on the plot of area A_0 rather than to a cell placed at exactly the location at which it would be placed under successive bisections. Nevertheless, as discussed in the text and in Appendices B.2 and B.3, the model does not provide realistic predictions of patterns in macroecology.

B.2 A fractal model for the SAR

If the $\alpha(n_0, A_0)$ are scale invariant, then using Eqs 3.12 and the results of B.1, the SAR can be expressed as:

$$\bar{S}(A_i) = \sum_{species} [\alpha(n_0, A_0)]^i. \quad (B.2.1)$$

Will the SAR obtained from Eq. B.2.1 will be a power law? To explore this, let's introduce another parameter, this time a community-level probability a_i, defined to be the ratio of the average number of species found at scale $i-1$ to that at scale i (Harte et al., 1999a). In terms of this parameter, we can write:

$$\bar{S}(A_i) = S_0 \prod_{j=1}^{i} a_j. \tag{B.2.2}$$

If the a_i are scale-independent, then Eq. B.2.2 reads, in analogy with Eq. B.1.4:

$$\bar{S}(A_i) = S_0 (a)^i. \tag{B.2.3}$$

Recalling that $A_i = A_0/2^i$, this is mathematically equivalent to the power-law form of the SAR:

$$\bar{S}(A_i) \sim A_i^z, \tag{B.2.4}$$

with

$$z = -\log_2(a). \tag{B.2.5}$$

A necessary and sufficient condition for a power-law SAR is that the a_i are independent of scale and thus all equal (Ostling et al., 2004). Now we have two expressions for the SAR: one based on the assumption of species-level scale invariance for all species (Eq. B.2.1), and one based on the assumption of community-level scale invariance (Eq. B.2.3). Are they equivalent?

If we assume scale-invariance at both species level and community level, and equate the expressions for $\bar{S}(A_i)$, we arrive at an important rigorously true impossibility theorem (Harte et al., 2001) relating the probability parameters at the species level and the community level:

$$a^i = <[\alpha(n_0, A_0)]^i>_{\text{species}}, \tag{B.2.6}$$

where $\langle \rangle_{\text{species}}$ denotes an average over all species.

I use the phrase "impossibility theorem" because Eq. B.2.6 cannot, in general, be satisfied for all i-values or even over an adjacent pair of i-values, $(i, i+1)$. Hence scale-invariance cannot hold simulataneously at the species level and the community level. The only case in which Eq. B.2.6 can be satisfied, and thus scale-invariance can hold at both community level and at species level, is if the αs are all equal to each other and at the same time equal to a. But that is impossible unless all species have the same abundance (Harte et al., 2001).

To summarize, if the αs are scale-invariant, then we arrive at a power-law range–area relationship for each species (Eq. B.1.4), whereas if the community parameter a is scale-invariant, then we arrive at a power-law SAR (Eq. B.2.4). But both metrics cannot be power-law except in the case in which all species have the same abundance and thus equal values of α.

Which, if either, option does nature choose? A robust observation is that the αs are not all equal for each species in systems where Eq. B.1.4 has been tested (Green et al., 2003). If the α s are not all equal, then by the impossibility theorem, we should not expect to see power-law behavior both for all the species-level $B(A_i)$s and for the SAR. In fact, we rarely see it for either.

B.3 A fractal model of the species-level spatial abundance distribution

To predict $\Pi(n|A_i, n_0, A_0)$, the model requires a boundary condition at some smallest scale, $i=m$, at which there is either 0 or 1 individual within each cell; the value of m may differ from species to species. This is different from the type of boundary condition assumed in Section 3.3.1, where the abundance of the species at some largest scale, A_0, was assumed. In addition I make the strong assumption that the probability parameter $\alpha(n_0, A_0)$ contains all of the possible information about cross-cell abundance correlations.

I define the probability distribution, $Q_i(n)$, for $n \geq 1$, to be the probability that, if a species is present in A_i cell, then it has n individuals in that cell. By scaling up from that smallest scale, $i=m$, the following relationship can be derived (Harte et al., 1999a) relating the conditional spatial probability distribution at two adjacent scales, A_i and A_{i-1}:

$$Q_{i-1}(n) = 2(1-\alpha)Q_i(n) + (2\alpha - 1)\sum_{\eta=1}^{n-1} Q_i(\eta) \cdot Q_i(n-\eta), \qquad (B.3.1)$$

subject to the boundary condition $Q_m(n) = \delta_{n,1}$. The symbol $\delta_{n,1}$ is called the Kronecker delta function; it has the value 1 if $n=1$ and value 0 if $n \neq 1$.

Eq. B.3.1 is an example of a recursive equation. It has a simple interpretation: The left-hand side is the probability that there are $n > 0$ individuals in a cell of area A_{i-1}. The right-hand side is the sum of the probabilities of all possible ways in which there can be $n > 0$ individuals in a cell of area A_{i-1}. First, they may all be in just one half of A_{i-1}, and thus in one of the two A_i cells that comprise the A_{i-1} cell. The coefficient $(1-\alpha)$ on the right-hand side of the equation is the probability that if the species occupies a cell, then its individuals in that cell are all on only one specified half of the cell. The factor of 2 multiplying that coefficient means that the individuals can be in only one or the other half of the cell... which half is not specified. Multiplying by the probability of there being n individuals in A_i, given that the cell is occupied, $Q(n)$, that first term on the right-hand side tells us the probability that the cell A_{i-1} has n individuals and all n of them are in either the right-or left-half of that cell.

A second option is that there are individuals in both halves of the A_{i-1} cell. The coefficient $2\alpha-1$ in front of the second term is the probability that individuals occupying a cell are found in both halves of the cell. This can occur in many ways: 1 individual on one half, all the other individuals on the other half, and so forth, so the summation term gives the joint probability of each of these options, under that assumption we made of independence among occupancy probabilities. Note that from the form of Eq. B.3.2, if there is only 1 individual at scale A_{i-1}, then the second term on the right-hand side does not contribute. So Eq. B.3.2 is simply a way of doing the book-keeping of all the options for occupying a cell A_{i-1} with various combinations of occupancies of the two halves of that cell.

It can be shown (Harte et al., 1999a) that the solution to Eq. 4.36 has the property that the average abundance of the species over *occupied cells* of area $A_i = 2^{m-i} A_m$ is $(2\alpha)^{m-i}$. Because there are n_0 individuals in A_0, consistency is achieved if:

$$m \log(2\alpha) = \log(n_0). \qquad (B.3.2)$$

So for each species, either its value of m or its value of α, or both, depend upon its abundance n_0.

For any particular species, the solution to Eq. B.3.1, at any fixed scale, i, and for $n \geq 1$, is unimodal, with a rising power-law dependence of the $Q_i(n)$ on n for small n, and a faster-than-exponential fall off at large n. Empirical data do not support this prediction (see Section 3.4 and the discussion of the shape of $\Pi(n)$).

In the publication in which Eq. B.3.1 was first derived (Harte et al., 1999a), we replaced the species-level parameter, α, by the community-level parameter, a, and claimed that the community $Q(n)$ could be interpreted as the species–abundance distribution $\Phi(n)$. Implicit in that argument, though not recognized or stated in the publication, was the assumption that the αs for all species are identical and equal to the community parameter, A. The assumption that species have identical α, however, does not hold for any actual ecosystem. Moreover, the solution to Eq. B.3.1 with a determined by Eq. B.2.5 of Appendix B.2 yields a very poor fit to typical species abundance distributions (Green et al., 2003).

Appendices B.1, B.2, B.3 together lead us to the conclusion that the fractal models proposed to explain species-level or community-level macroecological metrics unambiguously fail. The collection of models based on the common assumption of scale-invariant probability paramters α and a, could be called a theory because together they predict a wide range of phenomena, are parsimonious, and are falsifiable. Unfortunately, application of that last property has allowed us to conclude that the theory should be rejected.

B.4 A fractal model of community-level commonality

Consider two censused patches within the landscape, one of area A with $S(A)$ species ($S(A) = c\,A^z$) and one of area A' with $S(A')$ species. We assume, for now, that the distance D between the centers of the two patches is very large compared to either $(A)^{1/2}$ or $(A')^{1/2}$. We denote by $N(A, A', D)$ the number of species in common to the two patches, and use a similar notation now for X because it depends upon both areas. Commonality, defined as the number of species in common, divided by the average number of species in the two patches, $[S(A) + S(A')]/2$, is then given by:

$$X(A, A', D) = 2N(A, A', D)/(S(A) + S(A')). \qquad (B.4.1)$$

Consider the probability that a species, picked at random from the patch of area A', is also found in the patch of area A a distance D away. That probability, which we denote by $G(A, A', D)$, is given by the number of species in common to the two patches divided by the number in A', or from Eq. B.4.1:

$$G(A, A', D) = N(A, A', D)/S(A'). \qquad (B.4.2)$$

Using Eq. B.4.1, we can eliminate $N(A, A', D)$ from this expression, to get:

$$G(A, A', D) = [S(A) + S(A')] X(A, A', D)/[2S(A')]. \qquad (B.4.3)$$

Now suppose we double the size of the patch of area A'. Each of the species in A' is still nearly exactly a distance D away from the center of the patch of area A; (because $D \gg (A')^{1/2}$) and so, if a species is selected at random from the patch of area $2A'$, the probability that it also lies in the patch of area A is unchanged. In other words, $G(A, A', D)$ is independent of A' and, in particular:

$$G(A, A, D) = G(A, 2A, D), \qquad (B.4.4)$$

where A is an arbitrary area. Using Eq. (B.4.1):

$$G(A, 2A, D) = [S(A) + S(2A)] X(A, 2A, D)/[2 S(2A)] \qquad (B.4.5)$$

and

$$G(A, A, D) = X(A, A, D). \qquad (B.4.6)$$

Eq.B.4.2 tells us that the expressions in Eqs B.4.3 and B.4.6 are equal, so:

$$X(A, 2A, D) = \frac{2S(2A)X(A, A, D)}{S(A) + S(2A)}. \qquad (B.4.7)$$

Now using the power-law SAR, we obtain:

$$X(A, 2A, D) = \frac{2 \cdot 2^z}{1^z + 2^z} X(A, A, D). \qquad (B.4.8)$$

Again considering $X(A, A', D)$, we can continue with this same reasoning. If we double the size of the patch of area A, then we increase the number of species in that patch and thus increase the probability that a species selected at random a distance

D away from that patch of area A is also found there. The number of species increased by a factor of 2^z and so the probability can plausibly be assumed to increase by that same factor, or:

$$G(2A, 2A, D) = 2^z G(A, 2A, D) = 2^z G(A, A, D). \qquad (B.4.9)$$

Again using Eq. B.4.6, $G(2A, 2A, D) = X(2A, 2A, D)$, and so:

$$X(2A, 2A, D) = 2^z X(A, A, D). \qquad (B.4.10)$$

Equation B.4.10 holds true when D and z remain constant, and the censused areas double. Thus, Eq. 4.47 generalizes readily to the result $X(A, A, D) \equiv X(A, D) = (A)^z f(z, D)$, where f is some function. The form of the function f is constrained, however, because self-similarity, and the fact that X is a dimensionless fraction, imply $X = X(A/D^2)$. Therefore we must have:

$$X(A, D) = (A/D^2)^z g(z). \qquad (B.4.11)$$

In Exercise 4.9 you will derive the explicit form of the function $g(z)$.

Appendix C

Predicting the SAR: An alternative approach

Here we start with the METE results for the spatial abundance distribution at scale A_0, and the spatial abundance distribution $\Pi(n|A,n,A_0)$, and simply substitute into Eq. 3.13 to get the desired answer at any scale:

$$\bar{S}(A) = \bar{S}(A_0) \sum_{n=1}^{N_0} \frac{e^{-\beta_0 n}}{n \log(\beta_0^{-1})} (1 - \Pi(0|A, n, A_0)). \qquad (C.1)$$

The subscript 0 on β is to remind you that it is evaluated at scale A_0, and β_0 is given by the solution to Eq. 7.27 as a function of the values of $\bar{S}(A_0)$ and $N(A_0)$. $\Pi(0|A,n,A_0)$ in Eq. C.1 is determined from Eqs 7.48–7.50.

The predicted SAR can then be determined by numerical evaluation of Eq. C.1. For the case in which down-scaling of species' richness to determine $\bar{S}(A)$ is the goal, and the starting point is knowledge of $S(A_0)$ and $N(A_0)$, the procedure is straightforward. First solve for β_0 using Eq. 7.27 (or the simpler Eq. 7.30, if $\beta_0 N(A_0) \gg 1$ and $\beta_0 \ll 1$), solve for $\Pi(0)$ using Eqs 7.48–7.50, and then substitute that information in to Eq. 7.48 to determine $\bar{S}(A)$. The summation in Eq. C.1 can be carried out with many software packages, including Excel, MathCad, Matlab, and Mathematica.

To use Eq. C.1 to up-scale species' richness to scale A_0, starting with an empirical estimate of $\bar{S}(A)$, the procedure is more complicated because Eq. C.1 contains two unknowns, β_0 and $\bar{S}(A_0)$, and so does Eq. 7.27. Now, just as with method 1, the two equations have to be solved simultaneously, rather than sequentially.

It might be objected that Eq. C.1 really contains three unknowns, β_0, $\bar{S}(A_0)$, and $N(A_0)$, with only two equations (C.1 and 7.27) to determine them. As before, however, up-scaling abundance N from scale A to A_0 is not a problem. Because we are examining the complete nested SAR, an average of the measured N-values in each of the plots of area A is an estimate of the average density of individuals in A_0. Hence, we can assume:

$$N(A_0) = \frac{A_0}{A} N(A). \qquad (C.2)$$

C.1 A useful approximation for method 2

While Eqs 7.27, 7.48–7.50, and C.1 can be solved numerically to either up-scale or down-scale species' richness, it is useful to examine some simplifying approximations that will often be applicable to real datasets. With these simplifications we can obtain analytically tractable solutions that do not require numerical solutions and that provide considerable insight.

We assume that $A \ll A_0$, so that $1 - \Pi(0)$ can be approximated by Eq. 7.54. In practice, this means that we are considering ratios $A_0/A > 32$. Particularly for up-scaling applications, this assumption is not very constraining because we are often interested in up-scaling to areas much larger than the area of censused plots. As always, we assume $S_0 \gg 1$ so that $\exp(-\beta_0 N(A_0)) \ll 1$.

With $A \ll A_0$, we can replace Eq. C.1 with:

$$\bar{S}(A) \approx \bar{S}(A_0) \sum_{n=1}^{N(A_0)} \frac{e^{-\beta_0 n}}{n \log(\beta_0^{-1})} \left(\frac{n}{n + \frac{A_0}{A}}\right) = \bar{S}(A_0) \sum_{n=1}^{N(A_0)} \frac{e^{-\beta_0 n}}{\log(\beta_0^{-1})} \left(\frac{1}{n + \frac{A_0}{A}}\right). \quad (C.3)$$

Now we consider two cases separately: $\beta_0 A_0/A \ll 1$ and $\beta_0 A_0/A \sim 1$ (we will show that the case $\beta_0 A_0/A \gg 1$ is unlikely to arise).

Box C.1 shows that if $\beta_0 A_0/A \ll 1$, the summation in Eq. 7.63 results in:

$$\bar{S}(A) \approx \bar{S}(A_0)\left(1 - \frac{\log\left(\frac{A_0}{A}\right) + \gamma}{\log(\beta_0^{-1})}\right), \quad (C.4)$$

where γ is Euler's constant, 0.5772....

If $\beta_0 A_0/A$ is ~ 1, then a more complicated result, also shown in Box C.1, is obtained.

The SAR given in Eq. C.4 can be applied either to upscaling or down-scaling species' richness. As before, for down-scaling, Eq. 7.27 is used to determine β_0, which is then substituted into Eq. C.4 to determine $\bar{S}(A)$. For up-scaling, Eqs C.4 and 7.27 must be solved simultaneously. We see from the form of Eq. C.4 that for $A_0 \gg A$ and $\beta_0 A_0/A \ll 1$, the SAR is of the form:

$$\frac{\bar{S}(A)}{\bar{S}(A_0)} = a \log\left(\frac{A}{A_0}\right) + b, \quad (C.11)$$

where:

$$a = 1/\log(\beta_0^{-1}) \quad (C.12)$$

and

Box C.1 Derivation of Eq. C.4

We start with Eq. C.3 and let $A_0/A \equiv m_0$. Then, making the change of variable $m = n + m_0$, we can re-express Eq. C.3 as:

$$\bar{S}(A) \approx \frac{\bar{S}(A_0)}{\log(\beta_0^{-1})} \sum_{m=m_0+1}^{N_0+m_0} \frac{e^{-\beta_0 m} e^{\beta_0 m_0}}{m} = \frac{e^{\beta_0 m_0} \bar{S}(A_0)}{\log(\beta_0^{-1})} \sum_{m=1}^{N_0+m_0} \frac{e^{-\beta_0 m}}{m} - \frac{e^{\beta_0 m_0} \bar{S}(A_0)}{\log(\beta_0^{-1})} \sum_{m=1}^{m_0} \frac{e^{-\beta_0 m}}{m}.$$

(C.5)

Now recall the three assumptions that are being made here: $\beta_0 m_0 \ll 1$, $A_0/A \gg 1$, and $\exp(-\beta_0 N_0) \ll 1$. These assumptions, combined with the normalization condition on $\Phi(n)$, inform us that:

$$\frac{1}{\log(\beta_0^{-1})} \sum_{m=1}^{N_0+m_0} \frac{e^{-\beta_0 m}}{m} \approx 1.$$

(C.6)

Moreover, $\beta_0 m_0 \ll 1$ allows us to approximate $\exp(-\beta_0 m_0) \approx 1$ and the second summation on the right side of Eq. C.5 as (Abramowitz and Stegen, 1972):

$$\sum_{m=1}^{m_0} \frac{e^{-\beta_0 m}}{m} \approx \log(m_0) + \gamma.$$

(C.7)

Eq. C.4 immediately follows.

If $\beta_0 m_0 \approx 1$, then Eq. C.7 is no longer a good approximation, but the following expression quite accurately approximates the summation:

$$\sum_{m=1}^{m_0} \frac{e^{-\beta_0 m}}{m} \approx .643 \log\left(\frac{(A_0/A)^{1.55}}{0.408 + (\beta_0 A_0/A)^{1.55}}\right).$$

(C.8)

Eq. C.5 is now replaced with:

$$\bar{S}(A) \approx \bar{S}(A_0) e^{\beta_0 m_0} \left(1 - \frac{.643 \log\left(\frac{(A_0/A)^{1.55}}{0.408+(\beta_0 A_0/A)^{1.55}}\right)}{\log(\beta_0^{-1})}\right).$$

(C.9)

A third option, $\beta_0 m_0 \gg 1$, will not arise provided $\bar{S}(A) \ll N(A)$. To see this, we write:

$$\beta_0 A_0 = \beta_0 \frac{N(A_0)}{N(A)} \approx \frac{\beta(A) N(A)}{N(A)} < 1.$$

(C.10)

The inequality in this equation arises because $\bar{S} \ll N$ implies $\beta \ll 1$.

A useful approximation for method 2 • 243

$$b = 1 - \gamma/\log(\beta_0^{-1}). \tag{C.13}$$

From Eqs C.11–C.13, we can extract further information about the shape of the species–area relationship. For the case of down-scaling, β_0 is a fixed parameter determined using Eq.7.27 from the values of $\bar{S}(A_0)$ and N_0 at scale A_0. Hence, $\bar{S}(A)$ depends logarithmically on area.

For up-scaling species' richness, we assume $\bar{S}(A)$ is known and express $\bar{S}(A_0)$, with $A_0 \gg A$, as:

$$\bar{S}(A_0) \approx \frac{\bar{S}(A)}{1 - \frac{\log(\frac{A_0}{A}) + \gamma}{\log(\beta_0^{-1})}}. \tag{C.14}$$

Using Eq. 7.27, this can be rewritten as:

$$\bar{S}(A_0) \approx \frac{\bar{S}(A)}{1 - \frac{\log(\frac{A_0}{A}) + \gamma}{\frac{\bar{S}(A_0)}{\beta_0 N(A_0)}}} = \frac{\bar{S}(A_0)\bar{S}(A)}{\bar{S}(A_0) - \beta_0 N(A_0)[\log(\frac{A_0}{A}) + \gamma]}. \tag{C.15}$$

Rearranging terms, we derive:

$$\bar{S}(A_0) \approx \bar{S}(A) + \gamma \beta_0 N(A_0) + \beta_0 N(A_0) \log\left(\frac{A_0}{A}\right). \tag{C.16}$$

At first glance, Eq. C.16 does not seem very helpful because the coefficient β_0 in front of the $\log(A_0)$ term itself depends on $\bar{S}(A_0)$. However, numerical evaluation of the up-scaling SAR from simultaneous solution of Eqs 7.27 and C.14 indicates that $\beta_0 N(A_0)$ is nearly exactly constant over spatial scales for which our assumption $A_0/A \gg 1$ holds. In particular, with each doubling of area, $A_0 \to 2A_0$, N doubles and β is approximately halved. At anchor scale, $\beta(A)N(A)$ is only slightly smaller than $\beta_0 N(A_0)$ because while N exactly doubles with each area doubling, β decreases by slightly less than a factor of two for areas A_0 only a little bigger than A. The bottom line is that the coefficient $\beta_0 N(A_0)$ in front of the $\log(A_0/A)$ term in Eq. C.16 is approximated by the value of βN at anchor scale. Numerical solutions to the method 2 up-scaling equations (7.27 and C.4) for a variety of anchor scale boundary conditions confirm the accuracy ot the approximations that lead to Eq. C.16.

References

Abramowitz, M. and Stegen, I. (1972). *Handbook of Mathematical Functions, with Formulas, Graphs, and Mathematical Tables*, pp. 228–37. Dover Books, New York.

Adler, P. B., White, E. P., Lauenroth, W., Kaufman, D., Rassweiler, A., and Rusak, J. (2005). Evidence for a general species-time-area relationship. *Ecology* **86**, 2032–9.

Akaike, H. (1974). A new look at the statistical model identification. *IEEE Transactions on Automatic Control* **19**(6), 716–23.

Allee, W., Emerson, A., Park, O., Park, T., and Schmidt, K. (1949). *Principles of Animal Ecology*. W. B. Saunders and Co., Philadelphia.

Arhennius, S. (1896). On the influence of carbonic acid in the air upon the temperature of the ground. *London, Edinburgh, and Dublin Philosophical Magazine and Journal of Science* **41**, 237–75.

Arhennius, O. (1921). Species and area. *Journal of Ecology* **9**, 95–9.

Azaele, S., Muneepeerakul, R., Rinaldo, A., and Rodriguez-Iturbe, I. (2010). Inferring plant ecosystem organization from species occurrences. *Journal of Theoretical Biology* [doi: 10.1016/j.jtbi.2009.09.026].

Banavar, J. and Maritan, A. (2007). *The Maximum Relative Entropy Principle* [http://arxiv.org/abs/cond-mat/0703622i].

Banavar, J., Moses, M., Brown, J. et al. (2010). A general basis for quarter-power scaling in animals. *Proceedings of the National Academy of Sciences* [doi/10.1073/pnas1009974107].

Beck, C. (2009). Generalized information and entropy measures in physics. *Contemporary Physics* **50**(4), 495–510.

Bell, G. (2001). Neutral macroecology. *Science* **293**, 2413–18.

Ben-Naim, A. (2006). The entropy of mixing assimilation: An information-theoretic perspective. *American Journal of Physics* **74**(12), 1126–35.

Berlow, E., Neutal, A., Cohen, J. et al. (2004). Interaction strength in food webs: issues and opportunities. *Journal of Animal Ecology* **73**, 585–98.

Bliss, C. I. and Fisher, R. A. (1953). Fitting the negative binomial distribution to biological data. *Biometrics* **9**, 176–200.

Brissaud, J. (2007). MaxEnt mechanics. *arXiv physics* /0701127.

Brown, J. (1995). *Macroecology*. University of Chicago Press, Chicago.

Brown, J., Gillooly, J., Allen, A., Savage, V., and West, G. (2004). Toward a metabolic theory of ecology. *Ecology* **85**(7), 1771–89.

Buck, B. and Macauley, V. (1991). *Maximum Entropy in Action*. Oxford University Press, Oxford UK.

Bulmer, M. (1974). On fitting the Poisson lognormal distribution to species-abundance data. *Bio-metrics* **30**, 101–10.

Carey, S., Harte, J., and del Moral, R. (2006). Effect of colonization and primary succession on the species-area relationship in disturbed systems. *Ecography* **29**, 866–72.

Carey, S., Ostling, A., Harte, J., and del Moral, R. (2007). Impact of Curve Construction and Community Dynamics on the Species-Time Relationship. *Ecology* **88**(9), 2145–53.

Chave, J. (2004). Neutral theory and community ecology. *Ecology letters* **7**(3), 241–53.

Chave, J. and Leigh, E. (2002). A spatially explicit neutral model of β-diversity in tropical forests. *Theoretical Population Biology* **62**, 153–68.

Chesson, P. (2000). Mechanisms of maintenance of species diversity. *Ann. Rev. Ecol. Syst.* **31**, 343–66.

Cho, W. and Judge, G. (2008). Recovering vote choice from partial incomplete data. *Journal of Data Science* **6**, 155–71.

Clark, J. S. (2009). Beyond neutral science. *Trends in Ecology and Evolution* **24**, 8–15.

Clark, J. S. and McLachlan, J. S. (2003). Stability of forest diversity. *Nature* **423**, 635–8.

Clauset, A. and Erwin, D. (2008). The evolution and distribution of species body size. *Science*, **321**, 399–401.

Clauset, A. and Redner, S. (2009). Evolutionary model of species body mass diversification. *Physical Review Letters* **102**, 038103.

Clauset, A., Schwab, D., and Redner, S. (2009a). How many species have mass M? *American Naturalist* **173**, 256–63.

Clauset, A., Shalizi, C., and Newman, M. E. J. (2009b). Power-law distributions in empirical data. *SIAM Review* **51**(4), 661–703.

Cohen, J., Briand, F., and Newman, C. (1990). *Community Food Webs: Data and Theory*. Springer-Verlag, New York.

Coleman, B. (1981). On random placement and species-area relations. *Journal of Mathematical Biosciences* **54**, 191–215.

Condit, R. (1998). *Tropical Forest Census Plots*. Springer-Verlag & R. G. Landes Co., Berlin.

Condit, R., Ashton, P. S., Baker, P. et al. (2000). Spatial patterns in the distribution of tropical tree species. *Science* **288**, 1414–18.

Condit, R., Pitman, N., Leigh, Jr et al. (2002). Beta- diversity in tropical forest trees, *Science* **295**: 666–9.

Conlisk, E., Bloxham, M., Conlisk, J., Enquist, B. and Harte, J. (2007a). A class of null models of spatial relationships. *Ecological Monographs* **77**(2), 269–84.

Conlisk, E., Conlisk, J., and Harte, J. (2007b). The impossibility of estimating a negative binomial clustering parameter from presence-absence data. *The American Naturalist* **170**(4), 571–4.

Conlisk, J., Conlisk, E., and Harte, J. (2009). Hubbell's local abundance distribution: Insights from a simple colonization rule. *Oikos* [doi: 10.1111/j.1600-0706.2009.17765.x].

Connell, J. H. (1971). On the role of natural enemies in preventing competitive exclusion in some marine animals and in rain forest trees. In P. J. Den Boer and G. Gradwell, eds. *Dynamics of Populations*, pp. 298–312. Proceedings of the Advanced Study Institute on Dynamics of Numbers in Populations, Centre for Agricultural Publishing and Documentation, Wageningen.

Damuth, J. (1981). Population density and body size in mammals. *Nature* **290**, 699–700.

Daniell, G. (1991). Of maps and monkeys: an introduction to the maximum entropy method. In B. Buck and V. Macauley, eds. *Maximum Entropy in Action*, pp. 1–18. Oxford University Press, Oxford, UK.

Darwin, C. (1859). *On the Origin of Species by means of Natural Selection, or the preservation of favored races in the struggle for life*. J. Murray, London.
Dengler, J. (2009). Which function describes the species-area relationship the best? A review and empirical evaluation. *Journal of Biogeography* **36**, 728–44.
Dewar, R., (2003). Information theory explanation of the fluctuation theorem, maximum entropy production and self-organized criticality in non-equilibrium stationary states. *Journal of Physics A, Math. Gen.* **36**, 631–41.
Dewar, R., (2005). Maximum entropy production and the fluctuation theorem. *Journal of Physics A, Math. Gen.* **38**, 371–81.
Dewar, R. and Porte, A. (2008). Statistical mechanics unifies different ecological patterns. *Journal of Theoretical Biology* **251**, 389–403.
Diggle, P. (1983). *Statistical Analysis of Spatial Point Patterns*. Academic Press, London.
Dobzhansky, T. (1964). Biology, molecular and organismic. *American Zoologist* 4, 443–52.
Dodds, P., Rothman, D., and Weitz, J. (2001). Re-examination of the "¾ law" of metabolism. *Journal of Theoretical Biology* **209**, 9–27.
Drakare, S, Lennon, J., and Hillebrand, H. (2006). The imprint of the geographical, evolutionary and ecological context on species-area relationships. *Ecology Letters* **9**, 215–27.
Dunne, J. (2006). The network structure of food webs. In M. Pascual and J. Dunne, eds. *Linking Structure to Dynamics in Food Webs*, pp. 27–86. Oxford University Press, Oxford, UK.
Dunne, J. A. (2009). Food webs. In R. Meyers, ed. *Encyclopedia of Complexity and Systems Science*, pp. 3661–82. Springer, New York.
Durrett, R. and Levin, S. (1996). Spatial models for species-area curves. *Journal of Theoretical. Biology* **179**, 119–27.
Elith, J. and Leathwick, J. (2009). Species distribution models: Ecological explanation and prediction across space and time. *Annual Review of Ecology and Systematics* **40**, 677–97.
Elith, J., Graham, C., Anderson, R. et al. (2006). Novel methods improve prediction of species' distributions from occurrence data. *Ecography* **29**, 129–51.
Enquist, B. and Niklas, K. (2001). Invariant scaling relations across tree-dominated communities. *Nature* **410**, 655–60.
Enquist, B. J., West, G. B., Charnov, E. L., and Brown, J. H. (1999). Allometric scaling of production and life history variation in vascular plants. *Nature* **401**, 907–11.
Etienne, R. S. and Olff, H. (2004a). How dispersal limitation shapes species—body size distributions in local communities. *American Naturalist* **163**, 69–83.
Etienne, R. S. and Olff, H. (2004b). A novel genealogical approach to neutral biodiversity theory. *Ecology Letters* **7**, 170–5.
Etienne, R. S., Alonso, D., and McKane, A. (2007). The zero-sum assumption in neutral biodiversity theory. *Journal of Theoretical Biology* **248**, 522–36.
Evans, M., Hastings, P., and Peacock, B. (1993). *Statistical Distributions*. Wiley Interscience, New York.
Fisher, R., Corbet A., and Williams, C. (1943). The relation between the number of species and the number of individuals in a random sample of an animal population. *Journal of Animal Ecology* **12**, 42–58.
Frank, S. (2009). The common patterns of nature. *Journal of Evolutionary Biology* [doi: 10.1111/j.1420-9101.2009.01775.x].

Gaston, K. and Blackburn, T. (2000). *Pattern and Process in Macroecology*. Blackwell Scientific, Oxford.
Gabaix, X. (1999). Zipf's law for cities: An explanation. *Quarterly Journal of Economics* **114**(3), 739–67.
Gause, G. F. (1934). *The Struggle for Existence*. Williams and Wilkins, Baltimore.
Gillooly, J. F., Charnov, E. West, G. B. Savage, V., and Brown, J. H. (2001). Effects of size and temperature on developmental time. *Nature* **417**, 70–3.
Gilpin, M. and Soule, M. (1986). Minimum viable populations: Processes of species extinctions. In M. Soule, ed. *Conservation Biology: The Science of Scarcity and Diversity*, pp. 19–34. Sinauer, Sunderland, MA.
Gleason, H. A. (1922). On the relation between species and area. *Ecology* **3**, 158–62.
Golan, A., Judge, G., and Miller, D. (1996). *Maximum Entropy Econometrics: Robust Estimation with Limited Data*. J. Wiley and Sons, New York.
Goldstein, H. (1957). *Classical Mechanics*. Addison Wesley, Reading MA.
Grandy, W. (1987). *Foundations of Statistical Mechanics. Vol: Equilibrium Theory*. D. Reidel Publishing Co., Dordrecht.
Grandy, W. (2008). *Entropy and the Time Evolution of Macroscopic Systems*. Oxford University Press, Oxford, UK.
Gray, J., Bjorgesaeter, A., and Ugland, K. (2006). On plotting species abundance distributions. *Journal of Animal Ecology* **75**, 752–6.
Green, J., Harte, J., and Ostling, A., (2003). Species richness, endemism, and abundance patterns: tests of two fractal models in a serpentine grassland. *Ecology Letters* **6**, 919–28.
Green, J. L., Holmes, A. J., Westoby, M. et al. (2004). Spatial scaling of microbial eukaryote diversity. *Nature* **430**, 135–8.
Grinstein, G. and Linsker, R. (2007). Comments on a derivation and application of the "maximum entropy production" principle. *Journal of Physics A: Mathematical and Theoretical* **40**, 9717–20.
Gruner, D. (2007). Geological age, ecosysgtem development, and local resources constraints on arthropd community structure in the Hawaiian Islands. *Biological Journal of the Linneaen Society* **90**, 551–70.
Gutenberg, B. and Richter, C. (1954). *Seismicity of the Earth and Associated Phenomena, 2nd Ed*. Princeton University Press, Princeton.
Hardesty, B., Hubbell, S., and Bermingham, E. (2006). Genetic evidence of frequent long-distance recruitment in a vertebrate-dispersed tree. *Ecology Letters* **9**(5), 516–25.
Harnik, P. (2009). Unveiling rare diversity by integrating museum, literature, and field data. *Paleobiology* **35**, 190–208.
Harte, J. (2006). Toward a mechanistic basis for a unified theory of spatial structure in ecological communities at multiple spatial scales. In D. Storch, P. Marquet, and J. Brown, eds. *Scaling Biodiversity*, pp. 101–26. Cambridge University Press, Cambridge, UK.
Harte, J., Kinzig, A., and Green, J. (1999a). Self similarity in the distribution and abundance of species. *Science* **284**, 334–6.
Harte, J., McCarthy, S., Taylor, K., Kinzig, A., and Fischer, M. (1999b). Estimating species-area relationships from plot to landscape scale using species spatial-turnover data. *Oikos* **86**, 45–54.
Harte, J., Blackburn, T., and Ostling, A. (2001). Self Similarity and the Relationship between Abundance and Range Size. *American Naturalist* **157**, 374–86.

Harte, J., Conlisk, E., Ostling, A., Green J., and Smith, A. (2005). A theory of spatial structure in ecological communities at multiple spatial scales. *Ecological Monographs* **75**(2), 179–97.

Harte, J., Zillio, T., Conlisk, E., and Smith, A. B. (2008). Maximum entropy and the state variable approach to macroecology. *Ecology* **89**, 2700–11.

Harte, J., Smith, A., and Storch, D. (2009). Biodiversity scales from plots to biomes with a universal species-area curve. *Ecology Letters* **12, 789–97.**

Hastings, A. (1980). Disturbance, coexistence, history, and competition for space. *Theoretical Population Biology* **18**, 363–73.

He, F. and Gaston, K. (2000). Estimating species abundance from occurrence. *The American Naturalist* **156**, 553–9.

He, F. and Legendre, P. (2002). Species diversity patterns derived from species area models. *Ecology* **83**, 1185–98.

Holling, C. S. (1973). Resilience and stability of ecological systems. *Annual Reviews of Ecology and Systematics* **4**, 1–23.

Horner-Devine, M., Lage, M., Hughes, J., and Bohannan, B. J. M. (2004). A taxa–area relationship for bacteria. *Nature* **432**, 750–3.

Hubbell, S. P. (2001). *The Unified Neutral Theory of Biodiversity and Biogeography*. Princeton University Press, Princeton, New Jersey, USA.

Hubbell, S. (2006). Neutral theory and the evolution of ecological equivalence, *Ecology* **87**(6), 1387–98.

Hubbell, S. P., Foster, R. B., O'Brien, S. et al. (1999). Light gaps, recruitment limitation and tree diversity in a neotropical forest. *Science* **283**, 554–7.

Hubbell, S. P., Condit, R., and Foster, R. B. (2005). Barro Colorado forest census plot data. [URL http://ctfs.si/edu/datasets/bci].

Hurtt, G. and Pacala, S. (1995). The consequences of recruitment limitation: Reconciling chance, history, and competitive differences between plants. *Journal of Theoretical Biology* **176**, 1–12.

Hutchinson, G. E. (1957). Concluding remarks. *Cold Spring Harbor Quart. Biol.* **22**, 415–27.

Hyatt, L, Rosenberg, M., Howard, T. et al. (2003). The distance dependence prediction of the Janzen-Connell hypothesis: a meta-analysis. *Oikos* **103**, 590–602.

Janzen, D. H. (1970). Herbivores and the number of tree species in tropical forests. *The American Naturalist* **104**, 501–28.

Jaynes, E. (1957a). Information theory and statistical mechanics: I. *Physical Review* **106**, 620–30.

Jaynes, E. (1957b). Information theory and statistical mechanics: II. *Physical Review* **108**, 171–91.

Jaynes, E. (1963). Information theory and statistical mechanics. In K. Ford, ed. *Brandeis Summer Institute 1962, Statistical Physics,* pp. 181–218. Benjamin, New York, NY.

Jaynes, E. (1968). Prior probabilities. *IEEE Transactions on Systems Science and Cybernetics* **4**(3), 227–40.

Jaynes, E. (1979). Where do we stand on maximum entropy. In R. Levine and M. Tribus, eds. *The Maximum Entropy Principle*, pp. 15–118. MIT Press, Cambridge, MA.

Jaynes, E. T. (1982). On the rationale of maximum entropy methods. *Proc. Instit. Elec. Electron. Eng.*, **70**, 939–52.

Jaynes, E. (2003). *Probability: The Logic of Science*. Cambridge University Press, Cambridge, UK.

Judge, G. G., Miller, D. J., and Cho, W. K. T. (2004). An information theoretic approach to ecological estimation and inference. In G. King, O. Rosen, and M. Tanner, eds. *Ecological Inference: New Methodological Strategies*, pp. 162–87. Cambridge University Press, Cambridge, UK.

Kempton, R. A. and Taylor, L. R. (1974). Diversity discriminants for the lepidoptera. *Journal of Animal Ecology* **43**, 381–99.

Khinchin, A. I. (1957). *Mathematical Foundations of Information Theory*. Dover, New York.

King, G. (1997). *A Solution to the Ecological Inference Problem: Reconstructing Individual Behavior from Aggregate Behavior*. Princeton University Press, Princeton, NJ.

Kinzig, A. and Harte, J. (2000). Implications of endemics-area relationships for estimates of species extinctions. *Ecology* **81**, 3305–11.

Kirchner, J. W. and Weil, A. (1998). No fractals in fossil extinction statistics. *Nature* **395**, 337–8.

Kleiber, M. (1932). Body size and metabolism. *Hilgardia* **6**, 315–32.

Kleidon, A. and Lorenz, R., eds (2005). *Non-Equilibrium Thermodynamics and the Production of Entropy*. Springer, Heidelberg.

Kleidon, A, Malhi, Y., and Cox, P. (2010). Maximum entropy production in environmental and ecological systems. *Philosophical Transactions of the Royal Society B.* **365**, 1297–302.

Kolokotrones, T., Savage, V., Deeds, E., and Fontana, W. (2010). Curvature in metabolic scaling. *Nature* **464**, 753–6.

Krebs, C. J. (1989). *Ecological Methodology*. New York: Harper and Row.

Krishnamani, R., Kumar, A., and Harte, J. (2004). Estimating species richness at large spatial scales using data from small discrete plots. *Ecography* **27**, 637–42.

Kunin, W. E. (1997). Sample shape, spatial scale and species counts: implications for reserve design. *Biological Conservation* **82**, 369–77.

Kunin, W. E. (1998). Extrapolating species abundance across spatial scales. *Science* **281**, 1513–15.

Kunin, W. E., Hartley, S., and Lennon, J. (2000). Scaling down: On the challenge of estimating abundance from occurrence patterns. *American Naturalist* **156**, 553–9.

Loehle, C. (2006). Species abundance distributions result from body size-energetics relationships. *Ecology* **87**, 2221–6.

Lomolino, M. V. (2001). The species-area relationship: New challenges for an old pattern. *Progress in Physical Geography* **25**, 1–21.

Lotka, A. J. (1956). *Elements of Mathematical Biology*, Dover Publications, Inc. New York.

Lyons, S. and Willig, M. (2002). Species richness, latitude and scale sensitivity. *Ecology* **83**, 47–58.

MacArthur, R. H. (1957). On the relative abundance of bird species. *Proceedings of the National Academy of Sciences* **43**, 293–5.

MacArthur, R. H. (1960). On the relative abundance of species. *American Naturalist* **94**, 25–36.

MacArthur, R. H. and Wilson, E. O. (1967). *The Theory of Island Biogeography*. Princeton Monographs in Population Biology, Princeton University Press.

Maddux, R. (2004). Self similarity and the species area relationship. *American Naturalist* **163**, 616–26.

Mandlebrot, B. (1982). *The Fractal Geometry of Nature*. W. H. Freeman, San Francisco.

Marquet, P., Navarrete, S. and Castilla, J. (1990). Scaling population density to body size in rocky intertidal communities. *Science* **250**, 1125–7.

Martinez, N. (1992). Constant connectance in community food webs. *The American Naturalist* **139**, 1208–18.

May, R. M. (1975). Patterns of species abundance and diversity. In M. L. Cody and J. M. Diamond eds. *Ecology and Evolution of Communities*, pp. 81–120. Harvard University Press, Cambridge, MA.

May, R. M. (1990). How many species? *Philosophical Transactions of the Royal Society of London B* **330**, 293–304.

May, M., Lawton, J., and Stork, N. (1995). Assessing extinction rates. In J. H. Lawton and R. M. May, eds. *Extinction Rates*, pp. 1–24. Oxford University Press, Oxford, UK.

McGill, B. (2003). A test of the unified neutral theory of biodiversity. *Nature* **422**, 881–5.

McGill, B. and Nekola, J. (2010). Mechanism in macroecology: AWOL or purloined letter? Towards a pragmatic view of mechanism. *Oikos* **119**, 591–603.

Muller-Landau, H. C. et al. (2006). Testing metabolic ecology theory for allometric scaling of tree size, growth, and mortality in tropical forests. *Ecology Letters* **9**, 575–88.

Murray, J. D. (2002). *Mathematical Biology: An Introduction*. Springer, New York.

Nee, S., Harvey, P. H., and May, R. M. (1991). Lifting the veil on abundance patterns. *Proceedings of the Royal Society of London B* **243**, 161–3.

Nekola, J. and White, P. (1999). The distance decay of similarity in biogeography and ecology. *Journal of Biogeography* **26**, 867–78.

Nekola, J., Sizling, A., Boyer, A., and Storch, D. (2008). Artifactions in the log transformation of species abundance distributions. *Folia Geobotanica* **43**, 259–68.

Ostling, A., Harte, J., Green, J., and Kinzig, A. (2003). A community-level fractal property produces power-law species-area relationships, *Oikos* **103**, 218–24.

Ostling, A., Harte, J., Green, J., and Kinzig, A. (2004). Self similarity, the power law form of the species-area relationship, and a probability rule: A reply to Maddux. *The American Naturalist* **163**, 627–33.

Paltridge, G. (1975). Global dynamics and climate: A system of minimum entropy exchange. *Quarterly Journal of the Royal Meteorological Society* **101**, 475–84.

Paltridge, G. (1978). The steady state format of global climate. *Quarterly Journal of the Royal Meteorological Society* **104**, 927–45.

Paltridge, G. (1979). Climate and thermodynamic systems of maximum dissipation. *Nature* **279**, 630–1.

Phillips, S. and Dudik, M. (2008). Modeling of species distributions with MaxEnt: new extensions and a comprehensive evaluation. *Ecography* **31**, 161–75.

Phillips, S., Dudik, M., and Schapire R. (2004). A maximum entropy approach to species distribution modeling, pp. 655–62. *Proceedings of the 21st International Conference on Machine Learning*, ACM Press, New York.

Phillips, S., Anderson, R., and Schapire, R. (2006). Maximum entropy modeling of species geographic distributions. *Ecological Modeling* **190**, 231–59.

Pielou, E. C. (1975). *Ecological Diversity*. Wiley International, NY.

Pimm, S. (1982). *Food Webs*. Chapman and Hall, London.

Plotkin, J. B. and Muller-Landau, H. C. (2002). Sampling the species composition of a landscape. *Ecology* **83**, 3344–56.

Plotkin, J., Potts, M., Leslie, N, Manokaran, N., LaFrankie, J., and Ashton, P. (2000). Species area curves, spatial aggregation, and habitat specialization in tropical forests. *Journal of Theoretical Biology*, **207**, 81–99.

Preston, F. (1948). The commonness, and rarity, of species. *Ecology* **84**, 549–62.

Preston, F. W. (1962). The canonical distribution of commonness and rarity: Part 1. *Ecology* **43**, 185–215.

Pueyo, S., He, F., and Zillio, T. (2007). The maximum entropy formalism and the idiosyncratic theory of biodiversity. *Ecology Letters* **10**(11), 1017–28.

Purves, D. and Turnbull, L. (2010). Different but equal: the implausible assumption at the heart of neutral theory. *Journal of Animal Ecology* [doi.10.1111/j.1365-2656.2010.01738.x].

Ramesh, B. R., Swaminath, M., Santoshgouda V. et al. (2010). Forest stand structure and composition in 96 sites along environmental gradients in the central Western Ghats of India. *Ecology* **91**, 3118.

Ripley, B. D. (1981). *Spatial Statistics*. Wiley, New York.

Ritchie, M. (2010). *Scale, Heterogeneity, and the Structure and Diversity of Ecological Communities*. Princeton University Press, Princeton, NJ.

Rodriguez, P. and Arita, H. (2004). Beta diversity and latitude in North American mammals. *Ecography* **27**, 547–56.

Rominger, A., Miller, T., and Collins, S. (2009). Relative contribution of neutral and niche-based processes to the structure of a desert grassland grasshopper community. *Oecologia* [doi 10.1007/s00442-009-1420-z].

Rosenzweig, M. L. (1995). *Species Diversity in Space and Time*. Cambridge University Press, Cambridge.

Rosindell, J. and Cornell, S. (2007). Species–area relationships from a spatially explicit neutral model in an infinite landscape. *Ecology Letters* **10**(7), 586–95.

Sandel, B. and Smith, A. B. (2009). Scale as a lurking factor: incorporating scale-dependence in experimental ecology. *Oikos* **118**, 1284–91.

Shannon, C. (1948). A mathematical theory of communication. *Bell System Technical Journal* **27**, 379–423.

Shannon, C. E. and Weaver, W. (1949). *The Mathematical Theory of Communication*. University of Illinois Press, Urbana IL.

Shipley, B., Ville, D., and Garnier, E. (2006). From plant traits to plant communities: a statistical mechanistic approach to biodiversity. *Science* **314**, 812–14.

Sivia, D. S. (1993). From dice to data analysis: maximum entropy and Bayesian methods. *Neutron News* **4**, 21–5.

Sole, R., Manrubia, S., Benton, M., and Bak, P. (1997). Self-similarity of extinction statistics in the fossil record. *Nature* **388**, 764–767.

Srinivasan, U., Martinez, N., Dunne, J., and Harte, J. (2007). Response of food webs across several trophic levels to realistic extinctions. *Ecology* **88**, 671–82.

Stanley, S. (1973). An explanation for Cope's rule. *Evolution* **27**, 1–26.

Sugihara, G. (1980). Minimal community structure: an explanation of species abundance patterns. *The American Naturalist* **116**, 770–87.

Sugihara, G., Schoenly, K., and Trombla, A. (1989). Scale invariance in food web properties. *Science* **245**, 48–52.

Taper, M. and Marquet, P. (1996). How do species really divide resources? *The American Naturalist* **147**, 1072–86.

Taylor, L. R. (1961). Aggregation, variance and the mean. *Nature* **189**, 732–5.

Thomas, C, Cameron, A., Green, R. et al. (2004). Extinction risk from climate change. *Nature* **427**, 145–8.

Tilman, D. (1994). Competition and biodiversity in spatially structured habitats. *Ecology* **75**, 2–16.
Tokeshi, M. (1993). Species abundance patterns and community structure. *Advances in Ecological Research* **24**, 111–86.
Volkov, I., Banavar, J. R., Hubbell, S. P., and Maritan, A. (2003). Neutral theory and relative species abundance in ecology. *Nature* **424**, 1035–7.
Volkov, I., Banavar, J. R., Maritan A., and Hubbell, S. P. (2004). The stability of forest biodiversity. *Nature* **427**, 696–9.
Volterra, V. (1926). Variazioni e fluttuazioni del numero d'individui in specie animali conviventi. *Mem. Acad. Lineci Roma*, **2**, 31–113.
West, G., Brown, J., and Enquist, B. (1997). A general model for the origin of allometric scaling laws in biology. *Nature* **413**, 628–31.
White, E., Adler, P., Lauenroth, W. et al. (2006). A comparison of the species-time relationship across ecosystems and taxonomic groups. *Oikos* **112**, 185–95.
White, E., Ernest, S., Kerkhoff, A., and Enquist, B. (2007). Relationships between body size and abundance in ecology. *Trends in Ecology and Evolution* **22**, 323–30.
Williams, R. and Martinez, N. (2000). Simple rules yield complex food webs. *Nature* **404**, 180–3.
Williams, R. (2010). Simple MaxEnt models explain food web degree distributions. *Theoretical Ecology* **3**, 45–52.
Williamson, M. and Gaston, K. (2005). The lognormal distribution is not an appropriate null hypothesis for the species-abundance distribution. *Journal of Animal Ecology* **74**, 409–22.
Wright, D. (1991). Correlations between incidence and abundance are expected by chance. *Journal of Biogeography* **18**, 463–6.
Yodzis, P. (1980). The connectance of real ecosystems. *Nature* **284**, 544–5.
Yodzis, P. (1984). The structure of assembled communities II. *Journal of Theoretical Biology* **107**, 115–26.
Zillio, T., Volkov, I., Banavar, J., Hubbell, S., and Maritan, A. (2005). Spatial scaling in model plant communities. *Physical Review Letters* [arXiv:q-bio/0508033v1 [q-bio.PE]].
Zipf, G. (1949). *Human Behavior and the Principle of Least Effort*. Addison Wesley, Cambridge, MA.

Index

abundance
  relationship to energy and mass 47–51, 110, 142–4
  estimation from sparse data 82–3
  inference from presence-absence data 202
  mass–abundance relationship (MAR) 47–51, 75–6, 110
  see also species-abundance distribution (SAD)
Allee effect 82
allometry 194
alpha diversity 47
Amazon 21, 74, 78, 99–100
  extinction estimation 79–80
Anza Borrego 179, 182
area of occupancy 36–7
arthropods 179, 186, 189, 195, 196, 201–2
assembly rule/process 90–1, 105, 110, 211

Barro Colorado Island (BCI) 70–3, 81, 179, 181, 182, 185, 187
Bayes' Law 34, 113, 128, 205, 216–17
beetles, Amazonia 78–9
Bernoulli process 91
beta diversity 47
bias 22
binomial distribution 45, 61–3, 88
  negative 95–7
biodiversity estimation 78–9
biological complexity 6–8
biome 32, 78–80, 154–5, 201
bipartite network 51–2
birds 179
  dispersal distribution 40
  geographic ranges 35–6
body size 41, 69, 77, 172–3
  distribution 109, 230
Boltzmann constant 119, 122
Boltzmann distribution 120

branching network 16, 99
broken stick model 106–7, 110
Bukit Timah 179, 187

Cauchy distribution 66
cell occupancy 160, 161, 184
census 28–9
  census data 34–6, 53, 80, 101, 111–12, 151, 177–79, 186, 192–3, 201, 203, 205–6, 210, 232
  censusing procedures 180
central limit theorem (CLT) 8, 102
Clairborne Bluff molluscs 179
climate change 80–2
climate envelope 133–4, 205
Cocoli data 187
colonization 108
Coleman model 12, 61–2, 87–9, 96, 113, 87–9, 170, 181, 184–5
collector's curve 43–5, 70, 71–3, 105, 168–9
  derivation 170–1
commonality
  community-level 46–7, 73–5, 213, 237–9
  species-level 37–8, 63–5, 89
  under self-similarity 100–1
community
  commonality 46–7, 73–5, 213, 237–9
  energy distribution 47
comprehensiveness, of a theory 4, 5
confidence intervals 111
conservation 201–2
consistency constraints 213–16
constant connectance rule 111
Cope's rule 109
cross-community scaling relationship (CCSR) 50, 51
curvature 63

Damuth rule 75, 160, 169–73, 210
Darwin 6–7, 78, 231
dependence relationship 32–3
dipole–dipole forces 10
disordered states 120–1
dispersal 40–1, 65–6
  inter-specific dispersal–abundance relationship 52
distance decay 37, 74–5
distinguishable individuals 87–9
distribution
  abundance 33–5, 41, 61–3, 66–7
  binomial 45, 61–3, 88
  body size 109, 230
  Boltzmann 120
  Cauchy 66
  dispersal 40–1, 65–6
  energy 39, 47, 109
  fat-tailed 66
  Gaussian (normal) 8, 65–6, 97
  lognormal 66–7, 102–4
  logseries 58–60, 67
  mass 109
  metabolic rate 152–3, 174–5
  negative binomial 95–7
  Poisson 12
  prior 126, 143, 211
  probability 21–2, 32–3, 142, 204
  rank 152–3
  Zipf 60–1
diversity
  alpha 47
  beta 47
  biodiversity estimation 78–9
  gamma 47
  species 7, 13, 107, 119, 201, 212
down-scaling 162–3

earthquakes 15
ecological inference problem 20–1
ecology 6–8
ecological processes 6
economics 134–5
ecosystem structure function 142–5, 146–51
elegant theories 4
endemics 46
endemics–area relationship (EAR) 46, 73, 167–8
  and extinction 80–2
  METE prediction 190–3
energy distribution 39, 47, 109
  abundance relationship 47–51, 110, 142–4
  community 47
  intra-specific 39
energy equivalence 75–6, 169–73, 175, 209–10
entropy 117–24
  alternative measures 127–8,186
  see also maximum entropy (MaxEnt)
evolution 6–7, 229
extent 178–80
extinction 7, 15, 27, 42, 106, 108–9, 189, 230
extinction estimation 79–82
  under habitat loss 202–3

falsifiability, theory and 3–4, 5
fat-tailed distribution 66
Fisher's alpha 208
foodwebs 51, 135, 210
  linkages 51–2, 76, 111, 210, 218
  models 110–11
  web cascades 82
fractal models
  community-level commonality 237–9
  range–area relationship 233–4
  species-level spatial abundance distribution 236–7
  species–area relationship (SAR) 234–6
fractals 15, 99

Galileo 6, 7
gamma diversity 47
Gaussian (normal) distribution 8, 65–6, 97
general circulation models (GCMs) 222
geographic range 35–6
  see also range–area relationship
global mass–density relationship (GMDR) 49–51, 75–6
Gothic Earthflow 179
grain size 178–80
Grand Unified Theory 5, 229–31
grasshoppers 179
growth rates 102

habitat loss 202–3
habitat–occurrence associations 203–6
Hawaii 186, 189, 195, 196
HEAP model 93–5, 159–60, 174, 214
Hubbell, Stephen 12
  neutral theory of ecology (NTE) 12–13, 16–17, 66–7, 89, 104–5, 208–9

ideal gas law 9–10, 17
image reconstruction 20
image resolution 133
indistinguishable individuals 89–93

individual mass distribution (IMD) 50, 51, 75
individuals' curve 103
inference 19
information 121
information entropy 121–9
  see also maximum entropy (MaxEnt)
information theory 117, 121–3
input–output tables 134–5
insects 20, 21, 78–9, 100
islands
  biogeography 43, 108–9
  colonization 108
  species–area relationship 42–4, 80–1
its from bits 231

Janzen–Connell effect 64, 65
Jaynes, E. 19, 117, 125–9, 133–6

Korup 179

La Planada 179
Lagrange Multipliers 22, 124–8, 146–8, 151, 223
Lambir 179
Laplace, Pierre Simon 90–1
  generalized Laplace model 92–3
  principle of indifference 90–1
  rule of succession 90, 159
latitudinal heat convection 220–1, 222
lean theories 4
linear binning 55
linkages 51–2, 76, 111, 173–4, 210, 218
local mass–density relationship (LMDR) 49–51
logic of inference 19, 22, 229
lognormal distribution 66–7, 102–4
logseries distribution 58–60, 67
log–log plots 53–5
Lotka–Volterra equations 11, 107
Luquillo 179, 188

macroecology 27, 136
  census 28–9
  patterns 52–5
  questions 27–8
macrostates 119–20
mainland SAR 43–4, 68–9
mass distribution 109
mass–abundance relationship (MAR) 47–51, 75–6, 110
mass–density relationship
  global (GMDR) 49–51, 75–6
  local (LMDR) 49–51

maximum entropy (MaxEnt) 19, 111, 123–9, 225
  examples 131–3
  failure to work 130–1
  uses of 133–6
  see also METE (maximum entropy theory of ecology)
maximum entropy production (MEP) 136, 219–25, 230
mechanism 8–11
mechanistic theories 8, 11–12
metabolic rate 48–51
  energy equivalence 169–73, 175
  mass relationship 15–16, 77–8
  minimum 142
  see also energy distribution
metabolic rate distribution 152–3, 174–5
  METE evaluation 193–5
metabolic scaling theory (MST) 16–17, 52, 77–8, 194, 209–10
METE (maximum entropy theory of ecology) 14, 141–75
  community-level species-abundance distribution 186–90
  dynamic METE development 218–25
  ecosystem structure function 142–5, 146–57
  endemics–area relationship 167–8, 190–3
  energy distribution 152–3, 174–5
  energy equivalence 169–73
  entities 141
  failure patterns 196–7
  predictions 162–75
  spatial correlation incorporation 213–17
  species-level spatial distributions 146, 157–62, 180–6
  species–area relationship 162–7, 190–3
  state variables 141
  structure 142–6
  see also maximum entropy (MaxEnt)
metrics 29–30, 31
  community-level 31, 32
  importance of 78–83
  species-level 31, 32
  see also specific metrics
microstates 119–20
minimum viable population 83
MST see metabolic scaling theory (MST)
Mudumalai 179

natural selection 6, 89–90
negative binomial distribution (NBD) 95–7
nested SAR 42, 44–5, 69, 80–1

nested SAR (*cont.*)
  model testing 111–12
networks 51, 135
  bipartite 51–2
  branching 16, 99
  linkages 51–2, 76, 173–4
neutral theory of ecology (NTE) 12–13, 16–17, 66–7, 89, 104–5, 208–9
neutrality 89–90
  argument for plausibility 208–9
niche-based models 105–7, 210
nitrogen 102
normal (Gaussian) distribution 8, 65–6, 97

O-ring measure 38–9, 41, 47, 64, 89, 213
  calculation 215, 216–17
occurrence–habitat associations 203–6
optimization 13–14, 16, 22

Paltridge model of latitudinal heat convection 220–1, 222
Panama 74
  Barro Colorado Island (BCI) 70–3, 81, 179, 181, 182, 185, 187
parsimony 4, 5
partition function 125, 146, 148, 158, 204
Pasoh 179
physics 6–8
Poisson cluster model 97
Poisson distribution 12
pollinator–plant network 51–2
power-law 15
  determination 53–5
  models 68, 97–101
presence-absence data 202
Preston theory 102–4
principle of indifference 90–1
principle of least action 136
prior distribution 126, 143, 211
prior knowledge expansion 20–2
probability
  conditional 32, 34, 142
  density 39–40, 55
  discrete 32–3, 55
  distribution 21–2, 32–3, 142, 204

quantum mechanics 9–10

random placement model (RPM) 12, 61–3, 87–9
  see also Coleman model
range–area relationship 35–7, 63
  fractal model 233–4
rank-variable graphs 55–61, 181

rarity 41, 154–5
reaction–diffusion models 109
relationship
  abundance–body size 211
  allometric 194
  dependence 32–3
  endemics–area (EAR) 46, 73, 167–8, 190–3
  energy–abundance 47–51, 110, 142–4
  mass–abundance (MAR) 47–51, 75–6, 110
  range–area 35–7, 63, 233–4
  species–area (SAR) 30, 41–5, 68–73, 111–12, 162–7, 190–3, 240–3
Renyi entropy 127–8
reproductive fitness maximization 13
Ripley's K-statistic 38–9, 213
  calculation 215, 216–17
rule of succession 90, 159

San Emilio 71, 72, 179
scale collapse 165, 190, 191
scale invariance 15, 98–100, 233–4
  commonality and 100–1
scaling 14–16
  cross-community scaling relationship (CCSR) 50, 51
  down-scaling 162–3
  metabolic scaling theory (MST) 16–17, 77–8
  power-law relationships 15
  up-scaling 162–3, 166, 192–3, 201–2
scaling up species richness 201–2
self-similarity 99–100, 233–4
  commonality and 100–1
serpentine grassland data 71–3, 179, 188, 232
Shannon entropy measure 186
shape dependence 36
Sherman data 182
Sinharaja 179
slope 68–9
Smithsonian Tropical Forest Science (STFS) plots 64–5, 67
South Africa 179
spatial correlations 213–17
  see also species-level spatial abundance distribution
spatial structure 37–8, 202
species diversity 201
species richness
  related to area 166–7
  scaling up 201–2
species-abundance distribution (SAD) 41, 66–7
  community-level 186–90
  graphical representation 55–61

METE prediction 153, 174
theories 102–7
see also abundance; species-level spatial abundance distribution
species-level commonality 37–8, 64–5, 89
species-level spatial abundance distribution 33–5, 61–3, 146, 157–62, 174, 213
  fractal model 236–7
  METE prediction 180–6
  see also abundance; species-abundance distribution (SAD)
species–area relationship (SAR) 30, 41–5, 68–73
  extinction estimation 80–2
  fractal model 234–6
  island 42–4, 80–1, 108–9
  latitude dependence 69
  mainland 43–4, 68–9
  METE prediction 190–3
  model testing 111–12
  nested 42–5, 69, 80–1
  power-law behavior 99–100
  prediction 162–7, 174, 240–3
  scale collapse 165, 174
  slope of 68–9
  triphasic 69
state variables 14, 141
  dynamic theory 219
statistical mechanics 9–10

statistical models 87–97
statistical theory 12
stochasticity 177
subalpine forest 179

theory 3–5
  criteria for 3–4
  evaluation 177–8
  simplicity importance 16–17
  types of 8–16
thermodynamics 9–10, 117–21
  entropy 117–21
Tjallis entropy measure 186
topological networks 20
trophic links 135, 210, 218
tropical forest 71, 74, 78
  Smithsonian Tropical Forest Science (STFS) plots 64–5, 67

unified theory 5, 229–31
units of analysis 30
up-scaling 162–3, 166, 192–3
  species richness 201–2

web cascades 82
Western Ghats, India 74, 79, 101, 179, 192–3
Wheeler, John Archibald 10

Yasuni 179

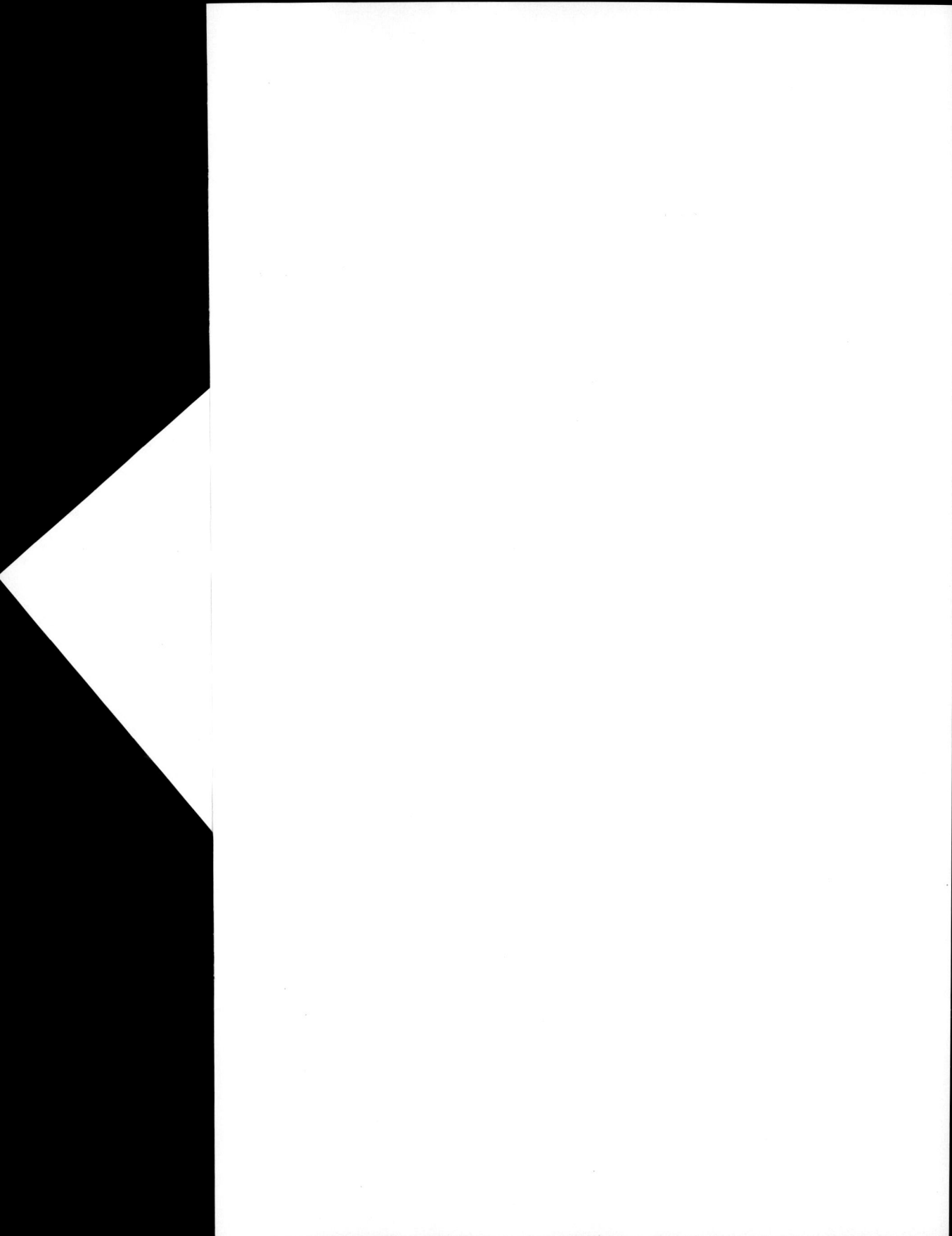

DATE DUE